WHISKY

WHISKY

AENEAS MACDONALD

BIRLINN

FIRST PUBLISHED IN MCMXXX
BY THE PORPOISE PRESS, EDINBURGH

THIS EDITION PUBLISHED IN MMXVI
BY BIRLINN
10 NEWINGTON ROAD, EDINBURGH EH9 1QS
www.birlinn.co.uk

PICTURE CREDITS

The author and publisher extend their grateful thanks to the following individuals and companies who have kindly provided period images for this edition. All copyrights are acknowledged. Facing page i Family Album of the late Mrs Anne Ettlinger. Facing page xviii, 52, 60, 65, 86, 91, 98, 111 Private Collection. Pages 6, 15, 71, 145 James & Linda Brown. Pages 7, 146 John Dewar & Sons Ltd. Page 33 The Kennetpans Trust. Pages 49, 95 Diageo plc Archive. Page 134—Old Matthew Giuseppe Begnoni, Whisky Paradise. Page 134—Abbot's Choice & Crawford's Sukhinder Singh Private Collection. Page 143 Morrison Bowmore Distillers Ltd.

A NOTE ON THE COPYRIGHT

George Malcolm Thomson lived until May 1996 and thus copyright on the text of *Whisky* extends until December 2066. On his death the copyright passed to his daughter Mrs Anne Ettlinger, who on her death in May 2014 bequeathed it to Ian Buxton, the present copyright holder. Copyright in all Thomson's other works remains with the family.

Main text, Appreciation, commentary and annotations
© Ian Buxton, 2016.

All rights reserved.

ISBN 978 1 78027 421 8

British Library Cataloguing-in-Publication Data
A catalogue record of this book is available on request from the British Library.

DESIGNED BY
Jules Akel

PRINTED BY
Asia Pacific Offset, Singapore

CONTENTS

PAGE

AENEAS MACDONALD'S *WHISKY*
—AN APPRECIATION ix

I. THE NATURE OF WHISKY 1

II. HISTORY 17

III. MAKING AND BLENDING 51

IV. GEOGRAPHY 79

V. JUDGING, PURCHASE, AND CARE 127

ACKNOWLEDGEMENTS & FURTHER READING 151

What sort of hivven's delight is this you've invented for all souls in glory?

C.E. Montague

> Sages their solemn een may steek,
> An' raise a philosophic reek,
> An' physically causes seek,
>   In clime an' season;
> But tell me whisky's name in Greek,
>   I'll tell the Reason.

Robert Burns

George Malcolm Thomson

AENEAS MACDONALD'S
WHISKY
AN APPRECIATION

Ian Buxton

A TIMELESS CLASSIC

I NEVER MET Aeneas MacDonald, though I could and should have. The author was in fact alive and well when I first encountered this little book, but as it had been published in 1930 and his name never appeared in print again I simply assumed—if I thought about it at all—that he was dead. That's what we all thought.

Secondly, there never really was an Aeneas MacDonald.

'But why should I read this book?' you ask. 'It's nearly ninety years old—why bother?'

It's not an unreasonable question: by a conservative estimate there are considerably more than two hundred books about, or mainly about, whisky, and new works are added at an ever increasing pace—a dozen or more in the last year alone. You may well feel that the world doesn't need any more books about whisky, and you could be right. Yet this was arguably the first. Doesn't that pique your curiosity?

Yes, despite the fact that the first written record of Scotch whisky may be found as long ago as 1494 and distilling is described in the sixth century by the Welsh bard Taliesin, apparently (extraordinary omission) no one thought it of sufficient interest or importance to its drinkers to write on their behalf until 1930 when this slim volume appeared. Whisky's slow renaissance began with its publication.

So for this alone Aeneas MacDonald deserves your attention, your respect and your time.

But there's more. MacDonald still speaks to us today: his sense of what it means to be Scottish; of why whisky, especially good whisky, matters; and on how, when and why to drink whisky he is a sure-footed and certain guide. And, at its simplest, this is also a damn good read: with glass and bottle at your side you will look with new enjoyment on your dram and dawning respect at a text that, yes, is approaching its ninetieth year but remains as fresh as a new-drawn cask sample: bright, sparkling and full of rich promise.

In fact, given today's near obsessive interest in the subject, and the acknowledged importance of whisky to the Scottish psyche, national identity and economy, it seems quite remarkable that we should have waited so long for MacDonald. As he says, 'whisky-drinkers are not yet mere soakers, in spite of the scant attention paid to their enlightenment by the trade' yet, until his work, the industry was a closed book to all but the insider. Thus, every subsequent writer on whisky, and every considered drinker, is in his debt.

THE MYSTERIOUS AUTHOR

But who was this Aeneas MacDonald? His name does not appear again and he would seem to have fallen silent after this single work, published initially by Edinburgh's Porpoise Press.

The answer is a curious one, for 'Aeneas MacDonald' was the pseudonym of a very much more prolific and long-lived Scot, who adopted a nom de plume for just this one volume. He was, in fact, George Malcolm Thomson, a Leither by birth (yet one who claimed Edinburgh as his spiritual home), Dux of Daniel Stewart's College and, after brief service as a 2nd Lieutenant in the Great War, a graduate of Edinburgh University.

Born in August 1899, Thomson had established the Porpoise Press in 1922 in partnership with his university friend Roderick Kerr and one John Gould, a sleeping partner whose investment of £5 made up the firm's initial capital of £20. Gould parted company with the Porpoise Press, apparently amicably, in 1925.

However limited its assets, its aims were rather more ambitious: 'a serious though modest effort to create a publishing house in Scotland of distinction for original work'. This bold manifesto was eagerly taken up by the Scottish press, the *Scots Pictorial* commenting favourably:

> Everyday it is being borne more forcibly upon our minds that there is a great need for a Scottish literary and publishing centre such as the founders of the Porpoise Press are endeavouring to establish. For there are young literary men in Scotland whose work is known only to a small circle, because there is no proper medium by which it may be introduced to the larger public which may be eager to know it.

By 1930 the Porpoise Press had issued some forty-eight titles, mostly original poems and novels by Scottish authors including George Blake, Eric Linklater, Hugh MacDiarmid and Neil Gunn. Thomson himself had contributed two titles at this point: the broadsheet *An Epistle to Roderick Watson Kerr* (as Kerr had left Edinburgh to take up a position on the *Liverpool Daily Post*) and a Porpoise Pamphlet titled *Will the Scottish Church Survive?*

Thomson, by then an active journalist and author, had also written two other more significant works: *Caledonia, or the Future of the Scots* and *The Rediscovery of Scotland*, both through the London publishers Kegan Paul, Trench, Trubner & Co. in 1928. The same publisher brought out his *A Short History of Scotland* in 1930.

WILL THE REAL GEORGE THOMSON, PLEASE, STAND UP?

The two earlier books on the economic condition of Scotland and his articles and broadcasts on BBC radio had attracted the attention of Lord Beaverbrook and, by 1930, Thomson had moved permanently to London to work first for the *Evening Standard* and subsequently across Beaverbrook's empire.

But in acknowledging the Thomson of the 1920s and 1930s as Aeneas MacDonald we are obliged to confront a problem: in both his early books on the Scottish economy and political situation he displays an ugly and bigoted attitude to Scotland's Irish-Catholic population that is blatantly racist. While nothing of this, mercifully, creeps into *Whisky* and, later in life, Thomson disavowed these views and claimed to regret them they have understandably and rightly sullied his reputation, and, so far as he is remembered at all in pre-war Scottish political life it is as the unacceptable face of a failed strain of proto-nationalism. It is a sorry legacy and a stain on his character that cannot be denied.

However, I believe that in the course of his long life we can discern three distinct George Malcolm Thomsons.

Before the war we find a passionate and angry young man, disenchanted by much in Scottish life and determined, through the creation of the Porpoise Press and, less admirably, through crude populist journalism of a deplorable kind, to change the direction of his country. At this time he was that strange beast: a nationalist who also favoured the Union with England and worked, sometimes overtly and sometimes behind the scenes, to influence the nascent movement for Scottish nationalism to accept home rule within the context of unionism. He also sought to improve Scotland's social and economic conditions and, seen in that light, much of his early work is prophetic and a valuable record of pre-war Scotland. But he was mistaken in his analysis of the impact of Irish-Catholic immigration and, even when this was pointed out by contemporaries, he held stubbornly to a prejudiced and intolerant set of attitudes.

APPRECIATION

By 1935 he appears to have matured and mellowed somewhat, perhaps influenced by life in London. His last pre-war book, *Scotland: That Distressed Area*, is more thoughtful, measured and balanced. It was well received at the time and went some way to restoring his reputation with contemporaries. If this work had been all he had written on the subject his name would be more highly regarded but, for all that he was not alone in his views on Irish immigration and Catholicism, the earlier bigotry remains an unsavoury episode and an uncomfortable reminder of a febrile time.

After the publication of this latter book his life moved to a new phase. Though he did not disagree with either Thomson's views or his analysis, Lord Beaverbrook was apparently unhappy about its publication, and his employer took the view that Thomson's talents should be reserved for the columns of his newspapers. Thomson was required to sign a new contract to that effect and from that date until he retired he did not publish anything in book form under his own name or any pseudonym.

Thomson enjoyed a long and successful career in journalism, interrupted only by work as one of Beaverbrook's closest advisers in various wartime ministries. He was regarded as one of Beaverbrook's closest, most long-standing and most trusted aides. Conducted largely in anonymity, this was an important part of his life and work but it was simply a precursor to the third phase of his work in which we see the mature Thomson working at full power until late in life.

His long retirement saw a burst of energy and creativity, with some dozen books, mostly well-regarded political commentaries and historical biographies and one novel, with a Jacobite theme, *The Ball at Glenkerran* (1982). After the passionate young Scot came the considered and professional journalist, carefully and discreetly serving his employer's wishes (and on occasion directing his thoughts). This Thomson was to be followed by the prolific and successful author and amateur historian, and perhaps this is the closest that we come to seeing the real Thomson. It is, I imagine, how he would like to be remembered.

Having modestly declined a decoration for his wartime service and after a distinguished career in journalism, George Malcolm Thomson was appointed OBE in 1990 for services to literature ('I thought they had forgotten me', he observed). He died in May 1996, aged ninety-six. A Scot through and through, he remained in London until his death.

APPRECIATION

WHO WAS AENEAS MACDONALD?

As a young man, Thomson studied both history and English literature at Edinburgh University. Perhaps it is here that we find the roots of his life-long fascination with Jacobitism and his strong sense of national identity. Certainly the nationalist tone of the Porpoise Press fitted with the growing resurgence of native pride and a distinctive sense of a Scottish identity, as well as the acceptance of youthful ideas prevalent in intellectual circles after the First World War, as the writers of the Kailyard were pushed aside in favour of an increasingly self-conscious Scottish Renaissance.

Christopher M. Grieve (Hugh MacDiarmid) abandoned self-publishing and Neil Gunn moved from Jonathan Cape to the Porpoise Press in acknowledgement of its overtly 'consciousness of Scottishness', as George Blake expressed the zeitgeist. George Malcolm Thomson's significance in all of this is for ever immortalised by Lewis Grassic Gibbon's dedication to Thomson of his novella Cloud Howe (the middle part of the Scots Quair trilogy).

But why adopt a pseudonym in the first place? The question is easily answered: Thomson's parents were strictly teetotal, and anonymity was necessary to avoid distressing his mother who felt particularly strongly on the subject as a consequence of her religious beliefs and work with the poor of Leith.

This, at least, is the explanation given by Alistair McCleery, who interviewed Thomson in August 1983 for a history of the Porpoise Press. Thomson himself, in private correspondence with Neil Gunn in November 1930, wrote:

> Please keep my name out of 'Whisky' if you don't mind…I have private reasons—and should hate people to think the P.P. existed to publish me alone.

Shortly afterwards he wrote again to Gunn:

> My mother is coming down & as she is an ardent T.T. (Good God!). I am having my little dinner on Thursday 22nd in the Norwegian Club where the wine is good and abundant. You had better meet my wife, child & mother for tea!!

('For tea' is heavily underscored in the manuscript.) Strangely, despite the evidence of the correspondence, his recent biographer George McKechnie thinks this unlikely. In fact there may have been several motives for his adoption of a pseudonym, including reaction to his earlier *Caledonia, or the Future of the Scots*. Published in 1928, this attracted a certain controversy, with Thomson described as 'the best-hated man in Scotland' (though this extravagant description may have owed something to an enthusiastic publisher aware of the value of controversy).

This book, and its immediate and more fully developed sequel, *The Re-Discovery of Scotland*, show Thomson disillusioned with Scotland and all its works. With their discredited and shameful attacks on Irish immigrants and the Catholic faith both are hard to read. *Caledonia*, indeed, is little more than a diatribe: passionate and forceful at its best, it is hard to avoid seeing in it the disappointment of a young man whose ambitious plans for the Porpoise Press had been dashed by the provincial Philistinism of dour Edinburgh (a city he found 'dominated socially by a caste of sport-talking lawyers, doctors and insurance officials'—*The Re-Discovery of Scotland*).

Certainly his thoughts were already turning to London by 1928 and, with sentiments such as '(Scotland) is a land of second-hand thoughts and second-rate minds', he was unlikely to win friends at home. There will have been many who heartily wished this brash young man on the first train to King's Cross.

Is it fanciful, therefore, to attribute a wish to avoid associating the opprobrium then attaching to the Thomson name to *Whisky* or, indeed, the Porpoise Press itself? These were sensitive times in the life of the venture: in recognition of the partners' burgeoning careers, largely out of Scotland, the house was being sold to Faber & Faber, and it may have been thought prudent to avoid courting unnecessary controversy.

Certainly this was a bitter pill for Thomson—much space in his two earlier books is given over to castigating a lack of domestic industry and the dominance of England in Scottish life. To have to relinquish the Porpoise Press to a London house must have seemed a betrayal.

However, as late as April 1930, he still harboured ambitions for the Porpoise Press, even under new ownership. Attempting to recruit Neil Gunn as an author, he wrote:

> the books will make Scotland take us seriously but the pamphlets … will fill a gap in a Scotland which has no serious press.

Be that as it may, the name Aeneas MacDonald would have meant more—indeed would have been richly symbolic—to Thomson's contemporaries, especially those with strong nationalist leanings. A pseudonym is not casually chosen. In adopting this identity Thomson tells us something of himself: a Jacobite, a believer in lost causes, a patriot and also a survivor.

For the original Aeneas MacDonald played a small but pivotal role in one of the most romanticised yet blood-soaked episodes in Scottish history: the '45,

APPRECIATION

a venture as ill-considered as it was daring. He was one of Bonnie Prince Charlie's 'Seven Men of Moidart'—the Parisian exiles who sailed from Nantes in July 1745 with Prince Charles Edward Stuart to raise his standard at Glenfinnan. More potent yet for this association, the Jacobite party first landed in Scotland on the island of Eriskay on 2 August 1745—Thomson's own birth date 154 years later and a link of which he was no doubt keenly aware from an early age.

MacDonald had two roles in the expedition: intending to go to Scotland on his own business affairs, he was persuaded to accompany Charles in order to win over his brother Donald of Kinlochmoidart and his many relatives and, more famously, he became the expedition's banker. However, despite Prince Charles Edward Stuart lodging at MacDonald's home in Paris, he was a reluctant convert to Charles' cause, and energetic pleading was required to gain his support.

Distrusted and insulted by the ferocious Highland clan chieftains, MacDonald nonetheless followed 'Charlie' to the bitter end. However, luckily for him, he did not stand at Culloden; expecting a consignment of Spanish gold he was in Barra to collect this as the Duke of Cumberland directed the butchery.

Later he was captured, and stood trial for High

Treason in Southwark in December 1747, where he was found guilty. The jury's recommendation for mercy being ignored, he was condemned to death but pardoned and released, whereupon he returned to France.

In dressing in Jacobite clothing then, Thomson consciously takes on an identity entirely sympathetic to his own Scottish Renaissance, and there are more than passing references to the '45 in MacDonald's text:

> Whisky found an honoured place in the commissariat of the Jacobite armies. It gave spirit and endurance to Montrose's men on those marvellous marches with which they confounded the hosts of the Covenant; it comforted the kilted soldiery on the mad memorable raid on Derby in 1745 and ministered to the Prince himself when he was a hunted man.

And, again, later in Chapter Two:

> Now dawned the heroic age of whisky, when it was hunted upon the mountains with a price on its head as if it were a Stuart prince…

Yet it is hard, indeed impossible, to reconcile this choice of identity with his virulently anti-Catholic writings of this period. It seems improbable that Thomson the propagandist and advocate of home rule within the Union would have readily accepted a Catholic absolutist monarchy. For me it marks the romantic in the man and his ability apparently to reconcile two essentially contradictory opinions.

STANDING ON THE SHOULDERS OF GIANTS: MACDONALD'S SOURCES AND INFLUENCES

Whisky has clearly been written at speed, and whisky itself is a metaphor for the wider condition of Scotland which, as we have seen, greatly concerned the young author. It is a journalist's book, and a whisky snob's book.

There is no real evidence that he visited distilleries or spoke to distillers and, of course, there was relatively little source material available to him. But he felt that whisky mattered and wanted to make this case as part of his wider commentary on Scotland. He evidently saw the decline in interest and knowledge of whisky as symptomatic of the wider decline of Scotland— 'the decline of whisky as a civilised pleasure is linked with the decay of taste in Scotland'.

As it turns out he was as wrong about the long-term future of whisky as he was about Irish-Catholicism. Ironically remembered for a book he would have considered slight, it's a further irony that the very blends he deplores eventually came to save the single malts that he celebrates so passionately. Without blending, many more distilleries would have been closed and their whisky lost for ever.

But he does exhibit some considerable knowledge of the industry, so where did he acquire this? A close reading of the text, and comparison with other earlier books (albeit for a specialist trade audience), leads me to suggest that he had two major sources, though strangely he does not seem particularly aware of Alfred Barnard's work *The Whisky Distilleries of the United Kingdom* (1887) which is so highly regarded today.

His major influences appear to have been T.F. Henderson's *Old-World Scotland; Glimpses of Its Modes and Manners* (1893) where he draws on the chapter on Usquebagh and more importantly J.A. Nettleton's 1913 work *The Manufacture of Whisky and Plain Spirit*. This is a lengthy and somewhat dry technical work but concludes with a thirty-page chapter 'What is Whisky?' which includes Nettleton's thoughts on the work and recommendations of the 1908/09 Royal Commission on Whiskey. It is well worth reading in its own right. Suffice to say here that Nettleton was not impressed, and MacDonald follows Nettleton's line of passionate disappointment very closely.

Professor George Saintsbury's *Notes on a Cellar-Book*

(1920) has also clearly been studied carefully, though a mere eight pages in a work of near two hundred are devoted to whisky (we devoutly wish for more from Saintsbury but he is silent). However, there is a tantalising link to this esteemed Grand Old Man of Letters: as the long-serving Professor of Rhetoric and English Literature at the University of Edinburgh, later Emeritus Professor, he was well known to (and may even have lectured) the two friends who founded the Porpoise Press.

Kerr and Thomson approached him, then aged eighty, in 1922 for an essay for the fledgling Porpoise Press but were declined. *Notes on a Cellar-Book* had already enjoyed some success—is it fanciful to imagine him serving his young admirers a glass, or even exciting their youthful imaginations with daring talk of a whole volume on whisky?

Certainly, MacDonald stands in awe of his old professor, describing him as 'patriarchal', one who has 'not succumbed to the degeneracy of the age', and quoting with reverence Saintsbury's views on grain whisky ('only good for blending, or for mere "drinkers for drunkee"'). Later, Saintsbury is repeatedly invoked as an authority beyond reproach.

The book begins and ends with with a quotation from *Another Temple Gone*, a short story by C.E. Montague. Though all but forgotten, Montague was sufficiently highly thought of in 1923 to be published by Chatto & Windus in a series alongside Lytton Strachey, Aldous Huxley, Arnold Bennett, Hilaire Belloc, A.A. Milne, Clive Bell and Roger Fry—illustrious company indeed and *Another Temple Gone* remains a delightful read.

Other influences are apparent throughout the text and, though there will no doubt be several that I have missed or failed to trace, I have attempted to illustrate these in the annotations.

THE CONDITION OF WHISKY IN 1930

In part, *Whisky* is a gloomy read, full of pessimism and the lost glories of a bygone age albeit one that existed more in MacDonald's imagination than historical fact. But what was the condition of whisky in 1930?

The date is all the more significant, nay extraordinary, when we consider the distressed state of the industry. The name of Pattison's still echoed ominously. This Leith and Edinburgh concern had driven ten other distillers into receivership, blackened the whole name of blending, and indirectly led to a Royal Commission on Whiskey following exposure at their criminal trial of the nefarious practices of their infamous blending vat.

APPRECIATION

The Pattison brothers were indeed part of 'A Big Boom' which for them was to end in bankruptcy and imprisonment.

The Pattison brothers may have been in jail by 1901, but their malign influence cast a long and dark shadow. Then, scarcely recovered from the Pattison's blending scandal and subsequent crash, Scotch whisky faced the triple challenges of the Great War, the Great Depression and Prohibition in the U.S.A., its most important market.

Between the beginning of the century and 1930, when this little book was published, some fifty single malt distilleries were closed, many permanently (save but one was opened). The industry in Campbeltown, once the most prosperous in the country, teetered on the brink of extinction. Whisky, today such an icon of national identity, was then in unparalleled crisis.

PUBLISHING HISTORY

Before turning to *Whisky*'s continued cultural significance and contemporary importance, book collectors may appreciate a note on its somewhat tangled publishing history.

There have been five editions, though only two are generally known.

The original is the first U.K. edition, published by the Porpoise Press of Edinburgh in 1930. As described above, MacDonald/Thomson was one of the founders of the Porpoise Press but, by 1930, it had been largely transferred to Faber & Faber and this first edition was published under the supervision of George Blake and Frank V. Morley of Faber & Faber. Printing was by Robert Maclehose & Co. at The University Press, Glasgow, and the first in a long succession of Porpoise Press titles to be produced there.

Whisky was published in an edition of 1,600—large by Porpoise Press standards, with a blue linen cover, paper title-slip on the spine and an attractive two-colour jacket. This, with the character affectionately known as the 'Cask Boy' or 'Tipsy' to his friends, was drawn by A.E. Taylor, who also contributed the distillery map. (Taylor also illustrated works by William Cobbett, A.A. Milne and others.)

The same character also appears, but in monotone, on the dust jacket of the 1934 first American printing, published by Duffield & Green of New York on 25th January that year. The cover is brown linen, with gold lettering stamped on the spine for title, author and publisher, together with a decorative thistle. While the dust jacket illustration is the same, the plates appear worn, and some of the finer detail has been lost; A.E. Taylor's initials and the date '30' have been removed from the illustration and the advertisements for other Scottish fiction on the rear of the jacket have been dropped.

This edition was also printed by Robert Maclehose & Co. at The University Press, Glasgow, and appears to be internally identical to the U.K. version, with the exception of the distillery map. This has been simplified and redrawn with many of A.E. Taylor's more artistic embellishments such as the fish, galleons in full sail and tiny sketches of bewhiskered kilted shepherds enjoying a dram removed. While much of the charm of the map is accordingly lost it must be conceded that it is now rather easier to read!

Judging by the comparative ease with which it is possible to find copies of the American edition, the print run was greater. With the ending of Prohibition in December 1933, interest in drinks in the U.S.A. was very high, and, despite the recessionary times, it seems that Duffield & Green were persuaded to risk a larger number of copies.

In 1953 Thomson declined a proposal from the Grove Press of New York for a new edition on the grounds that the text then required too extensive a revision. His identity remained a mystery.

However, there is a further and very rare third edition which is of particular interest to both whisky enthusiasts and collectors of American literature: the *Briefcase Breviary* (December 1930) by Henry & Longwell of Garden City, Long Island.

Frank Henry and Daniel Longwell were employees

of the publishers Doubleday, who had offices and a printing plant in Garden City. One of Doubleday's most renowned authors was Christopher Morley, eldest son of Frank Morley, an eminent Anglo-American Professor of Mathematics and the elder brother of Frank V. Morley, the director of Faber & Faber mentioned earlier.

Christopher Morley was a distinguished American journalist, novelist and poet with more than fifty books of poetry and novels to his credit, including the best-selling *Kitty Foyle* (1939), which was made into an Academy Award-winning film (Ginger Rogers for Best Actress; the film also enjoyed two further nominations). Known for his habit of carrying a briefcase stuffed full of books, Morley collaborated between 1928 and 1931 with Henry and Longwell to create their own imprint, designed to issue limited editions of slim volumes they considered interesting, largely for their friends. In mock salutation to Morley's briefcase, these were known as 'Briefcase Breviaries' and were published by the 'firm' of Henry & Longwell.

The edition is bound in yellow boards, with a device resembling a shark's fin in black repeated across the cover. Produced on high-quality laid paper, with a page size of 5½ by 8 inches, a total of 307 copies were printed. Each copy is individually hand-numbered and signed by the four protagonists immediately following the title page.

Whereas Henry, Longwell and Morley's signatures appear confident and flowing, that of Aeneas MacDonald is more tentative and formal. On all the copies I have verified the signature clearly reads 'McDonald'—a mistake unlikely to have been made by someone signing their own chosen alias and moreover a particularly improbable error for a passionate Scot. Did Thomson actually sign these copies or, as now seems more probable, did an American amanuensis perform this duty? If so, some minor literary subterfuge has just been uncovered.

I have been able to firmly locate fewer than thirty copies of this Henry & Longwell edition with certainty (it appears occasionally in the catalogues of American antiquarian booksellers, though the price has risen substantially in recent years). Delightfully, copy number one remains in Christopher Morley's library at the Humanities Research Center of the University of Texas at Austin. Of the surviving copies, many are inscribed by Morley to the recipient. This, and the fact that Morley, not the author, retained the first numbered copy speaks eloquently of Morley's role in this little adventure and his enthusiasm for the book itself.

His place in the story of *Whisky* does not end

there. In February 1934 in the *Saturday Review of Literature*, he 'puffed' the American launch in the following terms: 'Duffield & Green have published in full Aeneas MacDonald's admirable little panegyric on Whisky, one chapter of which was once privately issued under the furtive imprint of Henry and Longwell… "Aeneas MacDonald" is an alias of a well-known Scottish writer whose other works are of an ecclesiastical nature.'

The reference to MacDonald as an 'ecclesiastical' writer is an in-joke by Morley for those literary friends who shared the secret, a droll reference to Thomson's Porpoise Pamphlet *Will the Scottish Church Survive?* and the intense discussion of Scottish religious life in his earlier published works.

It conveniently also served to throw others off the scent—successfully, as the true authorship has never been widely known. Ingenious attributions include Archie MacDonnell, Scottish humourist and author of *England, Their England* (with its introduction by Christopher Morley). However, the link with Morley is a plausible connection, and, at the very least, the initials work and the putative author is a distinguished Scot of letters!

Morley's was not the only review. *The New York Times* of 4th February 1934 contained a long and generous notice by Edward M. Kingsbury (himself a 1926 Pulitzer prizewinner). It is headlined *'A Great, Potent, Princely Drink'* (alluding to MacDonald's opening paragraph) and subtitled *'Mr Aeneas MacDonald's Golden Treasury of Whisky Lore is Airy, Witty, Full of Sound Knowledge and Touched With Poetry'*.

Kingsbury quotes extensively from the text and, in rapturously praising the work, describes Aeneas MacDonald as follows:

> The author of this little poem, essay, history, geography, treatise and manual is admirably named as well as fitted for this task of love. The Highlands speak to him. Like his Roman predecessor, he brings back the old pieties and gods. There is something epic as well as georgic in his strain.

He goes on to applaud MacDonald's 'subtle imagination', commends many of MacDonald's remarks 'to our own distillers' (i.e. the American producers of bourbon and rye) and concludes: 'This is a small volume, but there are plenty of those who will love it. It is airy, witty, full of sound knowledge and practical wisdom.'

Similarly, George Currie reviewing for *The Brooklyn Daily Eagle* suggested that 'It should be savored leisurely, rather than gulped'. Sound advice.

Christopher Morley, Neil Gunn and Edward Kingsbury were not the only contemporaries to appreciate the book. The poet T.S. Eliot gave a copy to his close friend and mentor Harold Monro, of the Poetry Bookshop, inscribing it personally.

A lengthy hiatus followed the 1933 American edition until I was able to persuade Edinburgh's Canongate Books to issue a facsimile edition in October 2006. This fourth and final public edition is now out of print.

Finally, the distillers John Dewar & Sons produced their own strictly limited facsimile edition in 2012 for private circulation. However, the edition you now hold is the first to be illustrated with period material and provided with annotations intended to be helpful to the twenty-first century reader.

TODAY'S RELEVANCE AND APPEAL

We have established that contemporary reaction was favourable, even if further editions did not appear immediately. But what is this to us? The world of whisky is very different from 1930. Ownerships have changed along with distilling practice—today less than one-quarter of the Scotch whisky distilling industry remains in Scottish ownership, something that MacDonald would doubtless deplore and see as further evidence of Scotland's decline.

But, despite this gloomy statistic, whisky has staged a recovery that would have amazed MacDonald. Publications on whisky abound. What can Mr Aeneas MacDonald's Golden Treasury offer us today? Poetry, for one thing. Too many of today's whisky books are little more than lists: handsomely produced, well illustrated and comprehensive to a fault but with the soul of a draper's catalogue. Others might be mistaken for material straight from the distillers' own well-funded publicity machine, and a third category distributes marks out of a hundred to Glen This, Glen That and Glen The Other with the mechanical certainty of a drab provincial accountant.

Why any one 'expert' should be relied upon any more than the distiller's puff escapes me: in MacDonald's philosophy the enthusiastic drinker should learn and, thus informed, judge whisky for himself with 'his mother-wit, his nose and his palate to guide him'. Wise words.

We may list the faults in MacDonald, but he never once lacks for poetry. A love of whisky permeates his soul. Time is transcended by passages of lyrical beauty, and we are transfixed by the soaring spirit and graceful imagination of a true acolyte. Written in haste, sorrow and righteous anger, *Whisky* is simply a joy to read.

But we live in an age of league tables and cost-benefit analysis, where passion is distrusted and spreadsheets thought to contain the very secrets of our universe. Surely, in a world where an inventory of mash tun capacities passes for an initiation into the sacred mysteries of the still, MacDonald has nothing to teach us? What, you may ask, can I learn from this book?

Reader, take courage, for if utility is all you seek there is value still in these pages. Here, for example, we will find the first practical instruction in nosing and tasting whisky, MacDonald's guidelines for which, even including the selection of glassware, are reliable to this day.

In these pages, long-lost distilleries are brought to life and their merits discussed: the student will find much of value in the description of Islay and, even more poignant, Campbeltown whiskies. MacDonald's description of the Campbeltowns as 'the double basses of the whisky orchestra' is still quoted today and remains the classic descriptor of these 'potent, full-bodied, pungent whiskies'. How we now slaver for these lost glories! What the student of the cratur would not give to taste Rieclachan, Glenside or Benmore in their pomp!

He reminds us, too, that whisky has changed. 'The convenient proximity of a peat bog is an economic necessity for a Highland malt distillery' is not a sentence that could, in truth, be written today, and, in its very matter-of-factness, speaks volumes of the change in the taste of our drams. So, too, the thought that Highland malts will have the effect on men leading a sedentary life of making them liverish.

This particular myth—that malt whisky is too demanding for the effete office worker and demands the vigorous outdoor life of the Highlander of popular imagination—has passed into mythology. (We might note, en passant, that it sits curiously from one who lived by the pen for some seventy years.) It reappears as late as 1951 in Sir Robert Bruce Lockhart's *Scotch: The Whisky of Scotland in Fact and Story* and S.H. Hastie's *From Burn to Bottle* (published by the Scotch Whisky Association also 1951), and you may catch, from time to time, a faint echo repeated even today as a spurious endorsement of the 'manly' qualities of various single malts.

While this might seem today an outlandish conception, vigorously refuted both by the growing legions of female malt drinkers and the enthusiasm with which ever increasing levels of phenol are demanded in the newly fashionable malts of Islay, the thought must inevitably occur that perhaps our whisky has changed. Perhaps we are mere boys compared to MacDonald's heroic topers and—worrying thought—perhaps the price of whisky's world domination is that it has been rendered bland. These are deep matters to ponder, demanding the most pungent of drams to stimulate speculation.

However, while we may look with nostalgia on these long-lost peat-soaked monsters, elsewhere we find mention of some 128 blends comprising a mixture of Scotch and Irish spirit. This is a passing that cannot be lamented and, indeed, the ever reliable MacDonald trenchantly casts these headlong with two words—they are, he tells us, the 'crowning horror' of blending.

Clearly no apologist for the blender, in this as so much else, he anticipates our zeitgeist. MacDonald is a single-whisky man and a defender of the drinker against the conspiracy of silence that he describes as existing amongst blenders 'to prevent the consumer from knowing what he is drinking'. Well spoken, indeed.

His cry for clear labelling and precise descriptions is a modern one, and the industry could do worse than consider his plea that each label on a bottle ought to contain 'the names of the malt whiskies … in the blend, and the exact percentage of grain spirit … it should state the number of years and months that the blend and each of its constituents has matured in cask'.

Is this so unreasonable? As he notes: 'sound whiskies would only gain by it.'

A modest enough proposal, yet, I have no difficulty in prophesying, one that will be ignored before it is stoutly resisted 'by a body of men so assured of their commercial acumen as those who comprise the whisky trade'!

'Transparency' is today's watchword, and the current campaign by one smaller producer for openness on labelling is one that MacDonald would sympathise with and support with energy and feeling.

But strict E.U. regulations govern the marketing and promotion of spirits, effectively preventing the distiller from telling the consumer exactly what is in each bottle. As Compass Box Whisky say on their website:

> It turns out that Scotch whisky is one of the few products where it is prohibited by law to be fully open with consumers. This is an issue that affects every corner of the Scotch world (from Single Malt distillers to blenders) and limits the ability of the producer to share pertinent information with their customers.
>
> We believe the current regulations should change. That Scotch whisky producers should have the freedom to offer their customers complete, unbiased and clear information on the age of every component used in their whiskies. That those customers have the right to know exactly what it is they're drinking.

APPRECIATION

Does this seem familiar? Read MacDonald on Judging, Purchase, and Care (Chapter V) and you will find exactly this argument: in effect nothing has changed since he wrote. Consumers are still pleading for the information he requested nearly ninety years ago, but at last some enlightened producers have lent their weight to the cause. Let us trust it will not be another ninety years before we are trusted with this simple information.

All this is excellent stuff and confirmation, if any is needed, that MacDonald remains pertinent today. Indeed, he also anticipates the fashion for cocktails, while disparaging the then fashionable mix of whisky and soda; his recommendations on blending your own whisky remain sound, and his championing of the drinker against the complacency of an industry then in frightening decline is both robust and relevant.

So, today the 'Sixteen Men of Tain' assume a symbolic importance out of all proportion to their numbers. But distilling did not go entirely unnoticed in Thomson's two trenchant pleas for Scotland. In *Caledonia* he observes that 'the whisky industry is in even worse plight, as a result of high taxation and American prohibition. In one important centre only one distillery out of seven is working.'

And in *The Re-Discovery of Scotland* his argument is backed up by cold statistics, chilling even today: 'In

1925, there were 124 distilleries working in Scotland; in 1926 there were 113. The export of whisky in 1926 was 800,000 proof gallons less than in 1925, and 1,856,000 less than in 1924.'

Enough. Today whisky is in robust health. 'Single whiskies', or malts as we now style them, take an ever larger share of the market. Established distilleries are expanding, and independent distillers are once again opening their doors as a burgeoning 'craft' sector innovates and explores ever more arcane aspects of our national drink. Aeneas MacDonald may rest easy.

To the extent that pioneers such as George Malcolm Thomson built the foundations of this present success we are ever in his debt. He lit a torch that has burned brightly ever since and still illuminates our faltering steps. His is truly a great, potent and princely voice that will not be stilled.

To think that he was frightened of his own mother!

Ian Buxton
August 2016

I

THE NATURE OF WHISKY

Of the history, geography, literature, philosophy, morals, use and abuse, praise and scorn of whisky volumes might be written.[1] They will not be written by me. Yet it is opportune that a voice be raised in defence of this great, potent, and princely drink where so many speak to slight and defame, and where so many glasses are emptied foolishly and irreverently in ignorance of the true qualities of the liquid and in contempt of its proper employment. For, if one might, for a trope's sake, alter the sex of this most male of beverages, one would say that there be many who take with them to the stews beauty and virtue which should command the grateful awe of men. Though, in truth, there is little of the marble idol of divinity about this swift and fiery spirit. It belongs to the alchemist's den and to the long nights shot with cold, flickering beams; it is compact of Druid spells and Sabbaths (of the witches and the Calvinists); its graces are not shameless, Latin, and

---

[1] *This may strike today's reader as prophetic. While in 1930 MacDonald's was a lone voice the 'volumes' that he anticipates have indeed been written, and continue to flow from the presses in ever increasing quantities. Whether any approach the fervour and poetry of this little book I shall leave it to you to determine.*

abundant, but have a sovereign austerity, whether the desert's or the north wind's; there are flavours in it, insinuating and remote, from mountain torrents and the scanty soil on moorland rocks and slanting, rare sun-shafts.

But of those who contemn it a word. We shall describe them and, according to their deserts, either bid them begone or stay and be instructed. For the enemies of whisky fall under several headings.

There has of late come into being a class of persons who have learnt of wine out of books and not out of bottles. They are as a rule to be surprised drinking cheap champagne in secret but their talk is all of vintages and districts and *clos* and *châteaux*.

These dilettantes of the world of drinking are distinguished by weak stomachs and a plentiful store of snobbery. Wine merchants make of them an easy and legitimate prey. They are apt in quotation and historical anecdote, culling these from the books which honest men have written to advance the arts of civilization and to earn money. They roll great names on their tongues as though they were heralds marshalling the chivalry of France, or toadies numbering the peers they have fawned on.

In finding those qualities of bouquet and body which their textbooks bid them seek, they are infallible, provided the bottle has been correctly labelled.

They will, indeed, discover them before they have tasted the wine.

One drinks ill at their tables; it was in the house of such an one that an impolite guest remarked to his neighbour, 'This Barsac goes to my head like wine.'

These creatures have the insolence to despise whisky. Fresh from their conducted tour of the vineyards, the smellers of corks and gabblers of names sneer when its name is mentioned. It is, they declare, the drink of barbarians, offensive to the palate and nostrils of persons of taste; above all, it is not modish. For all that is southern and Mediterranean is in the mode among us. Civilization is a Latin word and culture comes by the Blue Train.[2] Better a rubber beach at Monte Carlo than all the sea-shores of the north. And, of course, we must affect enthusiasm for wine; it is so European, so picturesque and cultivated. It shows one has a certain background. A cellar is like a pedigree and requires less authentication. Nor is actual experience of bibbing necessary; a good memory and the correct books will suffice. If one is actually forced to drink, one can toy for a time with

---

[2] *The luxury Calais–Méditerranée Express night train was a watchword for fashionable travel between the wars. Known as the Blue Train (*Le Train Bleu*) after the colour of the sleeping cars, it features in novels by Agatha Christie and Georges Simenon. Wealthy playboy motorists such as Woolf Barnato and the 'Bentley Boys' famously participated in the Blue Train Races of the 1920s and 1930s. The service has now been superseded by the TGV.*

the glass in one's palm, discuss the merits of the wine, quote from Brillat-Savarin[3] and tell that anecdote about the Duc de Sully, open a learned debate on a possible incompatibility of temperament between the wine and the food with which it is proposed to wed it, and, when all else fails, confess to a delicacy of palate which the grosser forms of indulgence would outrage. By such stratagems are the absurd and unmanly inner weaknesses of the bookish wine snob concealed from the ridicule they deserve.

But let no man think that I would have the sublime impertinence to dispraise wine or attempt to compare things incomparable, and assuredly I do not criticize honest and discriminating lovers of wine who have learnt by original research and from the contents and not the labels of bottles. There is no quarrel between us. They are not to be found among the white-livered ranks of the traducers of whisky, whose excellences they will proclaim even when in the arms of their mistress.

I pass on to another type of enemy, the men who drink whisky. With pain and not without a hope that they may yet be saved, let us number their sins. Foremost among these is that they drink not for the

---

3 Jean Anthelme Brillat-Savarin (1755–1826), a French lawyer, politician, epicure and gourmet whose book The Physiology of Taste *(1825) laid the foundations of all subsequent writing on food.*

pleasure of drinking nor for any merits of flavour or bouquet which the whisky may possess but simply in order to obtain a certain physical effect. They regard whisky not as a beverage but as a drug, not as an end but as a means to an end. It is, indeed a heresy of the darker sort, doubly to be condemned in that it lends a sad, superficial plausibility to the sneers of the precious. Whisky suffers its worst insults at the hands of the swillers, the drinkers-to-get-drunk who have not organs of taste and smell in them but only gauges of alcoholic content, the boozers, the 'let's-have-a-spot' and 'make-it-a-quick-one' gentry, and all the rest who dwell in a darkness where there are no whiskies but only whisky—and, of course, soda.

Yet it may be unjust to lay on the souls of such false friends the burden of whisky's present lamentable plight when it has become less a delight to the mouths of men than an item in a pharmacopoeia, and is drunk for all sorts of illegitimate reasons, as by journalists[4] to quieten conscience, by the timid to avert catarrh, by inferior poets to whip up rhymes, and by commercial travellers to dull the memory of rebuffs. For those whose business it is to encourage the sale of whisky are not without guilt in the matter. They bandy

---

4 *MacDonald was, of course, a journalist—a profession widely noted for its abstemious manner and temperate habits. Perhaps he was having fun here at the expense of his colleagues.*

no words about the aesthetic aspect of their wares; their talk is all of moral and physiological advantages: 'It does you good'; 'After a hot tiring day'; 'Now that the cold weather is coming'; 'The doctor recommends it.'[5] This is to be deplored, for it plays directly into the hands of those who look on whisky at its lowest, as a mere brute stimulant, who believe that one bottle of whisky resembles another very much as one packet of Gold Flake cigarettes another packet of Gold Flake cigarettes, and whose nearest approach to discrimination is to say 'small Scotch (or Irish), please,' instead of the still more catholic 'a small whisky'.

As a result, there has been a tendency to abolish whisky from the table of the connoisseur to the saloon bar and the golf club smoke-room. The notion that we can possibly develop a palate for whisky is guaranteed to produce a smile of derision in any company except that of a few Scottish lairds, farmers, gamekeepers, and bailies, relics of a vanished age of gold when the vintages of the north had their students and lovers.

---

[5] *From the 1880s onwards much whisky advertising adopted strategies similar to those noted here. The infamous Pattison brothers even had a brand known as 'The Doctor'.*

MacDonald rightly 'deplored' the use of quasi-medical claims.
Dewar's advertised like so many others at the time but today no distiller
would consider this acceptable, even if it were not explicitly banned
in marketing codes the world over.

It may indeed be that the decline of whisky as a civilized pleasure is linked with the decay of taste in Scotland. For it was from Scotland that first England and then the world at large acquired the beverage, though the Scots must share with the Irish the honour of being its first manufacturers. While a high standard of culture was still to be found in a Caledonia less stern and wild than to-day and whisky still held its place in the cellars of the gentry and of men of letters, who selected it with as much care and knowledge as they gave to the stocking of their cellars with claret, whisky retained its place as one of the higher delights of mankind. But when Scots opinion was no longer to be trusted, the standards of the whole world suffered an instant decline, similar to that which would befall Burgundy if the solid bourgeoisie of France and Belgium were to perish suddenly and no one was left to drink the wine but the English and the Americans.

And at last a day dawned when it became possible for the average Englishman to be ignorant of the very names of all but four or five blends of whisky (all of which stared at him in enormous letters from hundreds of hoardings) and for the average English public-house to stock no more than three or four proprietary brands. It is small wonder if, in the face of an indifference and promiscuity so widespread, the distilling firms relaxed their standards: no restaurant would trouble to lay in *Cabinett-wein* if its clients asked only for hock and would pay only one price whatever was brought to the table. Yet, at the same time, one cannot quite exculpate the whisky-makers from the guilt of having chosen the easier and more profitable path at the expense of the prestige of their own commodity. It was their duty to guard it jealously, for it cannot be maintained seriously that those who make the food and the drink of man are in the same category as mere commercial manufacturers. Theirs is a sacred, almost a priestly responsibility, which they cannot barter away for turnovers and dividends without betraying their trust as custodians of civilization. Flour and beef, wine, brandy, and whisky are not in the same category as Ford cars or safety pins. Or ought not to be.

It may be admitted, too, that there is some excuse for those who fall into the error of the whisky-swillers, who drink it because it has an infallible result, a loosening of the tongue, a dulling of the memory, a heightening of the temperature, or what not. For whisky—even inferior whisky—has a potency and a directness in the encounter which proclaims its sublime rank. It does not linger to toy with the senses, it does not seep through the body to the brain; it communicates through no intermediary with the core of a man, with the roots of his consciousness; it speaks from deep to deep. This quality of spiritual instancy derives from the physical nature of the liquid. Whisky is a re-incarnation; it is made by a sublimation of coarse and heavy barley malt; the spirit leaves that earthly body, disappears, and by a lovely metempsychosis returns to the world in the form of a liquid exquisitely pure and impersonal. And thence whisky acquires that lightness and power which is so dangerous to the unwary, so delightful to those who use it with reverence and propriety.

In the eyes of the uninstructed, whisky is a mere mitigator of the rigours of northern climates and northern theologies. This is an inaccurate or, at least, an incomplete view. The precise effects of whisky on the human organism, though they differ widely from those of wine, are not less complex. It is less adept in bathing with a rosy hue men and the works of this world; it is not a protector of comfortable illusions. We do not find whisky drinkers discovering causes to love mankind which are not apparent to them when they cease to be under the influence of the spirit. They do not indulge in unreasonably optimistic visions of the beauty and perfection of things, nor do they sentimentalize over the supposed gaiety of a departed age. They do not bang mugs on the table and roar about the jolly Middle Ages; they are not impressed by the well-padded spiritual comforts of Catholicism. Some might say that whisky is a Protestant drink, but it is rather a rationalistic, metaphysical and dialectical drink. It stimulates speculation and nourishes lucidity. One may sing on it but one is at least as likely to argue. Split hairs and schisms flourish in its depths; hierarchies and authority go down before the sovereignty of a heightened and irresistible intuition. It is the mother's milk of destructive criticism and the begetter of great abstractions; it is disposed to find a meaning—or at least a debate—in art and letters, rather than to enjoy or to appreciate; it is the champion of the deductive method and the sworn foe of pragmatism; it is Socratic, drives to logical conclusions, has a horror of established and useful falsehoods, is discourteous to irrelevances, possesses an acuteness of vision which marshals the complexities and the hesitations of life into two opposing hosts, divides the greys of the world rigidly into black and white. Foolishly regarded as the friend of democracy, it is as much the scorner of democratic fictions as of servility, snobbery, plutocratic stupidity and aristocratic arrogance. It is the banner of the free human spirit which refuses to be chained to the political expedients it is compelled to devise. 'Whisky and freedom gang thegither,'[6] but the freedom is not a social myth or a national egotism but rather the profounder autonomy of the individual soul.

The English,[7] who, by definition, will believe anything that the Scots tell them, took whisky from their northern neighbours. It is doubtful, all the same, if they have ever taken to it. The ale-sodden Saxon has a temperamental inability to comprehend the true inner nature of whisky, yet there is no doubt that it has given him, more or less against his will, a harshness and an edge alien to his original nature. It has given him an Empire and the hobnailed livers of his proconsuls. Sometimes he may have wondered suspiciously what it really was, this warlock liquor that came to him out of the mists. How could the Scots, a disagreeable people cringing under the tyranny of the pulpits, ever have come to think of this liberating and audacious drink? The Scots have perhaps been misunderstood by the rest of mankind.

---

6 *The reference is to Robert Burns' poem* The Author's Earnest Cry and Prayer *(1786).*

7 *I would hardly bother to observe that this is a piece of tongue-in-cheek fun at the expense of MacDonald's English friends and colleagues were it not that in August 2006 the then editor of* Whisky Magazine *ran a furious editorial entirely missing this point. So, for the avoidance of doubt, this and much of the next page and a half is to be read as humorous—though by the time we reach MacDonald's list of the 'gifts of Scotland' a somewhat bitter and sardonic tone has crept in, reflecting his disillusionment with the general state of Scotland.*

They are, for example, the most anti-clerical people in the world. For what is the Presbyterian system but a drastic putting of the clergy in its place by the laity, a usurpation of church government by the unordained? They have a deep but erratic strain of frivolity; they have never taken themselves seriously as a nation; their patriotism is of an impatient and sporadic character. No one has ever doubted their courage but none of the captains who led them to battle could ever be sure whether they would run away at first sight of the enemy or remain and fight to the death. They were poor material for military adventurers to work with and they could never conceivably have had the lack of humour which was required to build the British Empire. The statute books of their old parliaments are full of laws against golf and football, for the Scots preferred these deplorably unwarlike sports to the practice of archery which would have fitted them for battle.

The gifts of Scotland to the world are strange indeed, if they come from a people serious to the point of gloom. What are they? Whisky, golf, porridge, haggis, the kilt, the bagpipes, music hall comedians, public-house jokes. A curious list of products for a staid and sober people!

But whisky now belongs not to the Scots but to the world at large. It has therefore to be examined in relation to mankind as a whole, that is to say to people who imperfectly understand its essential nature, who will regard it as a habit, mildly reprehensible or fortified by social sanctions, or as a medicine, who will not see in it an art or a science. The decline of whisky in our times may, therefore, be related to vagaries of the Zeitgeist more accurately than to commercial shortsightedness and the failure of an educated appreciation. This age fears fire and the grand manner, and whisky may have to wait for its apotheosis until there comes to life upon the globe a race of deeper daring which will find in the potent and fiery subtlety of the great spirit the beverage that meets its necessities.

But, in the meantime, what can be done by way of persuasion and exposition to rebuild the ruined altar of whisky, let us do.

The tyranny of the pulpits!

II

HISTORY

The origin of whisky is, as it ought to be, hidden in the clouds of mystery that veil the youth of the human race. Legend has been busy with it and has given it, like the Imperial family of Japan, an ancestor descended from heaven. Yet no Promethean larceny brought us this gentle fire; whisky, or rather the whole art and science of distillation, was the gift of the gods, of one god, Osiris. The higher criticism may hesitate to probe a myth so poetic, may be content to hold its peace and accept the evidence of those X's on casks which have been adduced as direct links with the Osirian mysteries. It is certain, at any rate, that distilling is venerable enough to have acquired a sacred character. Whether it was born in the Nile Valley or elsewhere is a question of no consequence. The Chinese knew it before the Christian era; arrack has been distilled from rice and sugar since 800 B.C.; distilling was probably known in India before the building of the great Pyramid; Captain Cook found stills in use among the Pacific islanders.

It has been held that those comparatively modern

peoples whom we used to call the ancients knew nothing of distilling, but this view overlooks a passage in the *Meteorology* of Aristotle in which the father of scientific method notes that 'sea water can be rendered potable by distillation; wine and other liquids can be submitted to the same process.' This seems at least to give brandy, and possibly whisky also, a place in the sun of Greek knowledge. But if lovers of whisky think to find a direct reference to their beverage in the fifth century Zosimus of Panopolis and his mention of the distillation of a panacea or divine water, they will make a mistake. Zosimus was not thinking of whisky. They may comfort themselves with the reflection that distillation from grain has normally preceded distillation from wine, as one might expect, the necessity of those who have no wine to drink being more insistent. The chances are, then, that whisky has a longer pedigree than brandy.

There is, in any case, a clearer reference to the brewing and distilling of grain liquors in the ancient world which opens the exciting prospect that Dionysos was the god of whisky before he was the god of wine. During his wars against the northern barbarians the emperor Julian the Apostate encountered and apparently tasted (and certainly disliked) a beverage made from barley. The occasion was celebrated in an epigram to this new or (as seems likelier) very old aspect of Dionysos[1] which is found in the *Palatine Anthology* (ix, 368) and has been translated as follows in Miss Jane Harrison's *Prolegomena to the Study of Greek Religion*:

TO WINE MADE OF BARLEY

Who and whence art thou, Dionyse? now, by
    the Bacchus true
Whom well I know, the son of Zeus, say: 'Who
    and what are you?'
He smells of nectar like a god, *you* smell of goats
    and spelt,
For lack of grapes from ears of grain your
    countryman, the Celt,
Made you. Your name's Demetrios, but
    never Dionyse,
Bromos, Oat-born, not Bromios, Fire-born from
    out the skies.

It used to be thought that the early Thracian god, Dionysos Bromios, was connected with the loud peal of thunder, but as Julian points out, he was more

---

[1] *MacDonald is here suggesting a Greek origin for the art of distillation. This whole passage on the origins of whisky is perhaps more romantic than historically accurate; more recent scholarship should be consulted if the reader is concerned with establishing whisky's genesis. Perhaps, as MacDonald suggests earlier, it is 'a question of no consequence' and his poetic approach contains a deeper truth than that revealed in any pedantic chronology.*

probably associated with a grain-made intoxicant of the northern tribesmen. If the deduction is correct, it would explain the name Demetrios sometimes applied to the god, for Demetrios means 'son of Demeter', the Corn-mother.

By what means the knowledge of distilling arrived at the twin nurseries of whisky from the East we know not. It is not absurd to conjecture that those Mediterranean trader-seamen who are known to the anthropologists under such a variety of names brought it with them when they came in search of tin and gold and the pearls of our rivers. It is—to shun the temptation of filling in the gaps of history—at least certain that distilling was known to the Celtic peoples at an early date. Taliesin, the Welsh bard of the sixth century, mentions the process in his Mead Song and hints at a widespread knowledge of it: 'Mead distilled I praise, its eulogy is everywhere.'

The scene shifts to Spain of the Moors and the twelfth century.[2] Abucasis has been named the first philosopher of the West who applied distilling to

---

[2] *MacDonald fails to remark on the link between early distilling and alchemy, and does not mention Jabir ibn Hayyan, also known as Geber, polymath and father of modern chemistry, despite his name appearing in at least one of his apparent sources. It also seems unlikely, but possibly he was ignorant of the later work of the German Hieronymus Brunschwig.*

More surprising still is his omission of the Wizard of Balwearie, Michael Scott, whose writings refer to 'aqua ardens', the earliest name for distilled alcohol. He returned to Scotland from study in Oxford, Paris and Toledo around 1230. Arguably he could be considered the father of Scotch whisky.

spirits. In the following century Arnoldus de Villa Nova, chemist and physician, disciple of the learned Arabs, describes distilled spirit and calls it the 'water of life'. Raymond Lully, chemist and philosopher of Majorca, praises the 'admirable essence', describing it enthusiastically as 'an emanation of the divinity—an element newly revealed to man, but hid from antiquity, because the human race was then too young to need this beverage, destined to revive the energies of modern decrepitude'. The very accents of whisky advertising in our own days are caught here; Lully was born six centuries too soon. But it turns out that Lully is not praising whisky. Let him describe the making of the elixir: 'Limpid and well-flavoured red or white wine is to be digested during twenty days in a close vessel by the heat of fermenting horse-dung, and to be then distilled in a sand-bath with a very gentle fire. The true water of life will come over in precious drops, which, being rectified by three or four successive distillations, will afford the wonderful quintessence of wine.'

Whatever the Arabs may have known, then, it is not proved that they anticipated the Scots and the Irish (or their Celtic ancestors) in the discovery of whisky. The whole problem is made more difficult by the fact that the early term for distilled spirit is sometimes used in Scottish and continental manuscripts for any kind of distilled spirit; sometimes there is no doubt whatever that whisky is 'the water of life' referred to and sometimes it is equally clear that brandy is indicated. There is a school which insists that *aqua vitae* is a corruption of *acquae vite*, 'water of the wine'. If this is indeed so, then it seems probable that in the earlier writers *aqua vitae* denotes a kind of brandy. But, on the other hand, the word 'whisky' is derived from the Gaelic '*uis gebeatha*', which is 'the water of life', a direct translation of the Latin *aqua vitae*.

Daring indeed is the man who will presume to take sides in a question of precedence between the Irish and the Scots. He is as likely to be damned for his interference by the one party as to be denounced for his partiality by the other. When therefore one records that whisky, authentic and indubitable, first enters history as a beverage in use among the Irish one hastens to add that this proves nothing against the Scots. The mere fact of the priority of written evidence may mean anything or nothing, and, in this case, the Scots have the powerful retort to throw at their rivals that the precedence is due only to the fact that the Irish were conquered earlier than the Scots and not less effectively. Soon after the invasion of Ireland in 1170 the English found the Irish making and drinking what they called *aqua vitae* but which is undoubtedly whisky. Probably it had been in use for centuries before in Scotland as well as in Ireland, for it is hard to believe that two peoples who had so much to do with one another in the way of fighting, trading, thieving, and combining to fight common enemies did not share with one another their most prized possession.

At any rate in Skene's *Four Ancient Books of Wales* it is stated that by the sixth century the Gaels were known as skilled distillers. *The Red Book of Ossory*, an Irish manuscript of the fifteenth century, contains a detailed Latin description of the method of distilling *aqua vitae* and its employments as a medicine, but as this spirit was made from wine at least a year old distilled ten times, we deduce that the Hibernian talent for distillation was not confined to barley malt. Stanihurst, writing in the reign of Queen Elizabeth, has a passage closely parallel to that of the Irish scribe: 'One Theoricus', he says, 'wrote a proper treatise of *aqua vitae*, wherein he praiseth it unto the ninth degree. He distinguisheth three sorts, simplex, composita, perfectissima.... Ulstadius also ascribeth thereto a singular praise, and would have it burn being kindled, which he taketh to be a token to know the goodness thereof. And trulie it is a sovereigne liquor if it be orderlie taken.'[3] In an early Latin

---

3 *A similar phrase appears in Raphael Holinshed's 1577* History of Ireland. *Appropriately enough, the Irish distillers John Jameson &*

Life of St Brigida we learn that the celebrated Celtic saint, whose name is still remembered in numberless wells in Scotland (and was there not a Bridewell in London?), re-performed the miracle of Cana, the water being transformed into a liquid more acceptable than wine to a whisky-drinking people:

'*Christi autem ancilla videns quia tunc illico non poterat invenire cerevisiam, aquam ad balneum portatam benedixerit et in optimam cerevisiam conversa est a Deo et abundanter sitientibus propinata est.*'

By the fifteenth century the manufacture and use of whisky was well established in the Scottish Highlands and its fame had already penetrated to the Lowlands and the court of the illustrious king James IV. The Scottish Exchequer Roll of 1494–1495 informs us that eight bolls of malt were delivered to one Friar John Car[4] to make *aqua vitae*.[5] In 1497 a

---

Sons, then of Dublin, took this as the title of a charming little promotional pamphlet issued in 1950.

[4] *It was, of course, Friar John Cor. It has been calculated that, with modern apparatus, around 1,500 bottles of whisky could be distilled from eight bolls (about 0.4 tonnes), though we do not know at what strength Cor would have distilled his spirit. Alternatively, other sources suggest that he would have produced around 130 litres of alcohol or 420 bottles in present-day terms.*

Cor distilled at the Tironensian abbey at Lindores in Fife, destroyed by followers of John Knox in 1559 and long left derelict. However, there are now plans to build a small distillery and visitor centre there.

[5] *There are some fifteen further references to* aqua vitae *in the Exchequer Rolls up to 1512; these may have been for experiments with gunpowder,*

gift of whisky was brought to the king at Dundee by a barber who got 9s. for his courtesy. Exactly a century later Fynes Moryson, the traveller, reported that there was a distinction between Scotch and Irish whisky, the distinction being, he thought, in Ireland's favour. But in fairness to Scotland, it may be doubted if Moryson had tasted the best of the Scotch spirit. The Highlanders were notoriously reluctant to part with their finer whiskies. At this time there were three kinds of whisky known in the Isles, graded by their quality and by the elaborateness of the distilling process: usquebaugh, trestarig, and usquebaugh-baul. But these were mysteries still guarded by the Grampians and the Western Seas.

Not that whisky was completely unknown in England. In Henry VIII's reign Irish immigrants opened distilleries there and about the same time the South acquired a liking for the Scottish variety of the spirit. Shakespeare may have been referring to some inferior imported whisky in *As You Like It*:

> For in my youth I never did apply
> Hot and rebellious liquors in my blood.

If this be so, the dramatist was only repeating by insinuation counsel which was incorporated into

---

cosmetics or even embalming fluid. We simply don't know.

actual law in the Hebrides in 1609. Nine years later, Taylor, the water poet, in describing a famous hunt in Mar, speaks of 'the most potent *aqua vitae*'.

Temperance legislation, which has pursued whisky and whisky's less intelligent admirers through the centuries, is first encountered in the fifth of the Statutes of Icolmkill[6] agreed to by the Island chiefs. This statute pointed out that one cause of the poverty of the Isles was the immoderate consumption of *aqua vitae*. It restricted its manufacture for sale and imposed severe penalties on law breakers. But it permitted every man to distil as much as he liked for his own use. Several years later the Scots Parliament passed an 'Act that nane send wynes to the Ilis', for the purpose of mitigating the deplorable events which are likely to occur when the grape is added to the malt of grain. There is evidence, too, for the belief that the Islesmen were not alone in discovering their whisky was intoxicating as well as delightful. The first specific mention of a man getting drunk on whisky occurred when Lord Darnley, the deplorable second husband of Mary Stuart, made one of his French

---

[6] *The reference is to the 1609 Statutes of Iona, the precise meanings of which are hotly debated. The effect of the statutes was to permit local chiefs to buy both wine and whisky from Lowland suppliers but to prohibit tenants and 'country people'—who were nonetheless permitted to make and consume their own spirits 'to serve thair awne housis'—from purchasing it from merchants. The aim seems to have been to improve the economic condition of the poor.*

friends drunk on *aqua composita*, a kind of whisky.

During the first half of the seventeenth century whisky's popularity grew in the Scottish Lowlands, especially in the west, which was in closer contact with the Highland distilleries. All along the Highland Line, where Saxon and Gael met, distilleries were springing up, making the spirit which was sold in Glasgow taverns. During the Restoration period, when deep drinking was as popular a form of reaction against an overdose of Puritanism in Scotland, as were wenching and play-going in England, whisky's fame spread. In the *Mercurius Caledonius* of 1661 there is an account of a race run by twelve browster wives, at the instigation of some young Edinburgh blades, from Portobello to the top of Arthur's Seat. The reward for the dame who finished first in this more than athletic test was a hundredweight of cheese and a budgell of Dunkeld *aqua vitae*. By this time whisky was being made in Morayshire, where there was a local variety known as 'burnt *aqua vitae*', and in Glasgow, as well as in Dunkeld and other strictly Highland areas. But the distilleries were small and they catered chiefly for local needs. Probably the finest whiskies were distilled by Highland chiefs for the use of their own households; the classic brands were still unknown.

As early as 1505 the Royal College of Surgeons

at Edinburgh were given the monopoly of distilling and selling *aqua vitae* within the City bounds. If they had preserved this right, there would have been small need for the Edinburgh medical faculty to collect fees from their patients and perhaps the prestige of the Edinburgh medical school would never have been acquired. The first news we get of a really famous whisky is in 1690 when the 'ancient brewery of aquavity' at Ferintosh, in the possession of the Whig Forbes of Culloden, was sacked by marauding Jacobite Highlanders and 'all the whiskie pits...destroyed'. In recompense Forbes, or Culloden as we should style him in the good Scottish territorial fashion, was given a 'tack' or rent of the excise of the land. In effect, he was allowed to distil his whisky, already noted for its quality, without paying the excise on the malt. In a year or two it was said that as much whisky was distilled at Ferintosh as in all the rest of Scotland, and Culloden was making so much money that the Master of Tarbet complained to the Scots Parliament. But it was only in 1784 that the government compounded with Culloden for the exemption at a cost of £21,000. By that time 'Ferintosh' was almost as frequently used to describe the spirit as 'whisky' itself, and Burns somewhat prematurely mourned the passing of its ancient glories in his *Scotch Drink*:

Thee Ferintosh! O sadly lost!
Scotland laments frae coast to coast!
Now colic-grips an' barkin' hoast,
    May kill us a';
For loyal Forbes' chartered boast
    Is ta'en awa![7]

An exciseman, writing in 1736 of the manners of the Highlanders, pays a tribute to the virtues of whisky: 'The ruddy complexion, nimbleness and strength of those people is not owing to water-drinking but to the *aqua vitae*, a malt spirit which is commonly used in that country, which serves both for victual and drink.'[8] Whisky found an honoured place in the commissariat of the Jacobite armies. It gave spirit and endurance to Montrose's men on those marvellous marches with which they confounded the hosts of the Covenant; it comforted the kilted soldiery on the mad memorable raid on

---

[7] *Though Burns was himself, towards the end of his short life, an Excise officer he was well aware of the unpopularity of that branch of government service and the widely held view that the taxation of malt and barley was a breach of the 1707 Treaty of Union. However, the Excise officers could laugh at themselves: Burns wrote his song 'The Deil's Awa Wi' The Exciseman' for his colleagues at a 1792 Excise court dinner, where doubtless it was performed with vigour.*

[8] *Regrettably, I have been unable to trace the source of this quotation though subsequent writers on whisky such as Robert Bruce Lockhart and Joseph Earl Dabney happily repeat it without attribution.*

Derby in 1745 and ministered to the Prince himself when he was a hunted man. He had already tasted it in Culloden House on the eve of the fatal battle. It is reported that on one occasion he and two others finished the bottle at a sitting, the Prince consuming more than half the contents. When the 'Highland Host' came to Ayrshire to plague the stubborn West-country Whigs, William Cleland[9] said that each man of theirs thought his flask of whisky as essential a part of his military equipment as his claymore:

> There's yet I have forgotten
> Which ye prefer to roast or sodden,
> Wine and wastle, I dare say,
> And that is routh of usequebay.

The verses are poor things; the observation was no doubt accurate.

Smollett says that 'The Highlanders regale themselves with whisky, a malt spirit as strong as Geneva, which they swallow in great quantities without any signs of inebriation.' That they had many opportunities for doing so, we gather from

---

9 *The chronology is somewhat confused and confusing here: Cleland wrote his mock poem sometime before 1697 as a satire on the perceived fondness of Highland Scots for whisky.*

The Lyon in Mourning[10] which snobbishly dismisses the hospitality of Skye, 'all the publick houses there being mere whiskie houses'. But it is doubtful if the Highlanders were so free from the consequences of over-indulgence in whisky as Smollett would have us believe. There is evidence that the crafty Hanoverians found a way of employing whisky to loosen Highland tongues and loyalties, as a letter written by Lieutenant-Colonel Watson from Fort Augustus in 1747 shows: 'How soon the posts are fixed the commanding officer at each station is to endeavour to ingratiate himself in the favour of some person in his neighbourhood by giving him a reward, or by filling him drunk with whisky as often as he may judge proper, which I'm confident is the only way to penetrate the secrets of these people.'

But whisky has a graver and more honourable association with the Jacobite romance than any that has yet been mentioned. On the crucial night before Culloden, when Prince Charles was drinking in Culloden House, other men were finding a strange and sacred use for whisky. In Bishop Forbes'

---

10 *In* The Lyon in Mourning, *a collection by Robert Forbes, Episcopal Bishop of Ross and Caithness, of reminiscences about the Jacobite Rising, we learn that Bonnie Prince Charlie would 'put the bottle of brandy or whiskie to his head and take his dram without any ceremony'. It is said that in later years Forbes permitted favoured guests to drink out of Prince Charlie's brogues which he retained as a souvenir.*

Journal we read: 'Mr. John Maitland, chyrurgeon for the Soule...a Presbyter of the Episcopal Church of Scotland...was attached to Lord Ogilvie's regiment in the service of Prince Charles, 1745. He administered the Holy Eucharist to Lord Strathallan on the Culloden field (where that gallant nobleman received his death wound), it is said, with oat cake and whisky, the requisite elements not being attainable.'

During the greater part of the eighteenth century whisky was, in the Lowlands, an entirely unfashionable drink. Yet here one must distinguish. The cheap, popular whisky, distilled by Robert Stein at Kilbagie, who was easily the largest manufacturer, was poor stuff, worth no more than the penny a gill it fetched.[11] It was this raw, fiery tipple that the

---

11 *MacDonald is being a trifle harsh here. James and John Stein were real pioneers and it can be argued that their distilleries at Kilbagie and nearby Kennetpans represent the crucible in which modern Scotch whisky was formed.*

Tax changes in 1786–88 led to the Steins being effectively blocked from the English market where much of their spirit (which was indeed produced rapidly in shallow stills but in truth designed for rectification into gin) was sold. Large quantities of this 'poor stuff' then flooded onto the local market. The Steins quickly went bankrupt, though they subsequently reopened.

The scale of their operations was very considerable even by today's standards and they were significant innovators in whisky production. Robert Stein pioneered an early form of continuous still (1826) at the Kirkliston distillery, near Edinburgh (now closed). An improved design was patented by Aeneas Coffey in 1831 and soon enjoyed widespread success.

Funds are now being raised by a charitable trust to stabilise and eventually

Part of the distillery complex at Kennetpans

claret-drinking Lowland gentry and professional people despised and this, no doubt, that the mob in Aberdeen tossed off in half mutchkins to celebrate the acquittal of Lord George Gordon in 1781. But the cultivated taste of the lairds and the merchants did not reject the finer whiskies—when they could be obtained. 'Whisky in these days,' says a writer on Glasgow clubs of the eighteenth century,[12] 'being chiefly drawn from the large flat-bottomed stills of Kilbaggie, Kermetpans, and Lochryan, was only fitted for the most vulgar and fire-loving palates; but when a little of the real stuff from Glenlivet or Arran could be got it was dispensed with as sparing a hand as curaçoa or benedictine.'

These were the days of great drinking in the North, when Smollett's Highland gentleman in *Humphrey Clinker* spoke for his class in thinking it a grave disparagement of his family that not above a hundred gallons of whisky were consumed at his

---

restore the few remaining buildings at Kennetpans—for more information search 'Kennetpans Trust' online.

[12] *The writer on Glasgow clubs is John Strang (1795–1863) in* Glasgow and Its Clubs *(1st edition, 1856), quoted in T.F. Henderson's* Old-World Scotland: Glimpses of Its Modes and Manners *(1893). That appears to have been a major source for MacDonald, notably in this 'History' section, as he draws freely on Henderson's chapters 'On Wine and Ale' and 'Usquebagh'. He will doubtless have had other sources, particularly Bremner's* Industries of Scotland *(discussed later) but Henderson's work is evidently a major influence.*

grandmother's funeral, and when an ancient retainer pleaded with the guests at his master's funeral, 'that it might never be said to the disgrace of the master's hospitality that any gentleman that was his guest got to bed otherwise than being carried'. Whisky and funerals appear to have had a particularly close connection,[13] for we have another account of a similar servant who barred the exit to the guests with the solemn words: 'It was the express will o' the dead that I should fill ye a' fou' and I maun fulfil the will o' the dead'.[14] Need it be said that the pious duty was performed to the letter? 'I assure you, Sir,' said Bozzy to his mighty patron and victim, 'there is a great deal of drunkenness in Scotland.' Whereupon the Doctor uttered this majestic and memorable tribute to Scottish doggedness and capacity: 'No, sir, there are people who died of dropsies which they contracted in trying to get drunk.' But that may merely be an expression of Johnson's contempt as

[13] *Indeed, whisky and funerals are closely linked in the literature. As a further example, in 1616 the funeral expenses of Sir Hugh Campbell of Calder are said to have amounted to nearly £1,650 Scots, of which fully one quarter was accounted for by whisky for the mourners.*

[14] *From* Notes and Sketches Illustrative of Northern Rural Life in the Eighteenth Century *by William Alexander, 1877. He is describing the funeral rites observed for Sir Alexander Ogilvy, Lord Forglen, a Judge of the Court of Session in 1727. The servant was Forglen's clerk, one David Reid. As the story ends, 'he did fulfil the will o' the dead, for before the end o' 't there was nae ane o' us able to bite his ain thoomb'.*

a port-loving Englishman for what was in these days the main beverage of the upper-class Scotsman. ('Claret for boys'.) Whisky must not be blamed for the intemperance of eighteenth century Scotland, when judges took their bottle of claret with them into court and when the enlightened opinion on the use of alcohol might be found in the profound remark of a Lord of the Court of Session (himself no whisky-drinker) in addressing a jury: 'It is said, gentlemen, that the accused perpetrated this atrocious deed when he was drunk. Gentlemen, if he would do this when he was drunk, what would he no' do when he was sober!'

Powerfully assisted by the taxes on malt and ale, the popularity of whisky spread in Lowland Scotland during the second half of the eighteenth century. Yet the whisky which thus conquered new territories was the rawest, cheapest, and nastiest of its kind, consumed because it produced the desired result of drunkenness more speedily and economically than ale. Unfortunately, too, when whisky found its way into literature after so many centuries of exile, it was praised by poets whose taste was unlikely to be discriminating. Robert Fergusson, the impoverished Edinburgh clerk, certainly did not drink himself into Bedlam on Glenlivet; Burns, the arch-propagandist of the spirit, was an Ayrshire peasant with a peasant's taste for what was fiery and instant.¹⁵ It is probable that the poet laureate of whisky had but the slightest acquaintance with a whisky pleasing to civilized palates and that his muse, as he calls it, was nothing but the abominable Kilbagie of his day. No word of Burns' gives the slightest impression that he had any interest in the mere bouquet or taste of what he drank; on the contrary, his eloquent praise is lavished on the heating, befuddling effects of whisky. If this was all that Burns sought, it would be idle

---

15 *MacDonald's attitude is surprising to contemporary taste, and his judgement seems to me harsh.*

In Robert Burns and Pastoral: Poetry and Improvement in Late Eighteenth-Century Scotland *(OUP, 2010) Professor Nigel Leask reassesses Burns' reputation and places him as arguably the most original poet writing in the British Isles between Pope and Blake, and the creator of the first modern vernacular style in British poetry. In particular, Leask places Burns' work in the context of the revolutionary transformation of Scottish agriculture and society in the decades between 1760 and 1800, thus setting it within the mainstream of the Scottish and European enlightenments. As opposed to MacDonald's dismissive view, we may see Burns' writing on whisky as both perceptive and politically challenging.*

Burns continues to be celebrated around the world, not least every January 25th, and a full appreciation of his global significance and the accessibility of much of his language is overdue. A majority of Scots have long understood this.

To characterise Burns' attitude to whisky as one of indiscriminate consumption careless of quality fails to take account of the incendiary political stance of many of the poems on whisky. Moreover, as Alex Kraaijeveld has illustrated, Burns' private correspondence with John 'Auchenbay' Tennant does on at least one occasion show a vivid appreciation of the qualities of good whisky.

to suppose that he had any real appreciation of the merits whisky might have. And Burns' attitude to whisky has, through the extraordinary popularity of his verse, become almost universal. Whisky has been sung into fame as a ploughman's dram.[16]

The taxes on native liquors and the duties on foreign spirits gave whisky such a swift popularity in the eighteenth century that it became in its turn an object of taxation. The earliest known excise of twopence per gallon was imposed in 1660; when English enthusiasm for the drink awakened, an import duty of half a crown a gallon on Scotch whisky was introduced. Immediately a great smuggling trade sprang up—as much as 300,000 gallons crossed the Border in one year. Then came the charging of a license duty on the contents of the still used by the distillers—ingeniously circumvented by the Scots, who altered their stills so as vastly to increase the speed of production.[17] The net result

---

[16] *Burns was far from alone in lamenting the state of whisky. Distilling had been suspended in 1795 due to two consecutive poor harvests. When production resumed an anonymous chapbook,* Cheap Whisky, *celebrated the change of policy, and the renowned Perthshire fiddler Niel Gow composed his jaunty Strathspey 'Welcome Whisky Back Again' which is played to this day.*

[17] *As so often in MacDonald there is an inconsistency apparent; here, between his professed admiration for the ingenuity of the Scots' distillers in increasing the speed of their production and his earlier description of the output of the self-same altered stills at Kilbagie as 'abominable' and 'poor stuff', a 'raw, fiery tipple'.*

of all this legislation was to create an immense illicit distilling industry in the Highlands. As late as 1820 at least half the whisky sold came from unlicensed distilleries. In 1814 all distillation in the Highlands in stills of less than 500 gallons was prohibited. But it was almost as impossible to make such legislation effective as it is to prevent a free American citizen from killing himself with the hip-flask horrors of the era of Prohibition. There were thousands of stills in the glens and the shielings, and not all of the whisky they made was bad. Much of it, on the contrary, was better than the products of the licensed distilleries, for it was made in parts of the country where the science of distilling, like the art of playing on the bag-pipes, was an immemorial tradition, and where the qualities of water and soil favoured it.

Now dawned the heroic age of whisky, when it was hunted upon the mountains with a price on its head as if it were a Stuart prince, when loyal and courageous men sheltered it in their humble cabins, when its lore was kept alive in secret like the tenets of a proscribed and persecuted religion. If whisky has not degenerated wholly into a vile thing in which no person of taste and discernment can possibly take an interest, it is because its tradition was preserved, by men whose names ungrateful posterity has forgotten, during years when the brutal and jealous Hanoverian government sought to suppress in the Highlands this last relic of the ancient Gaelic civilization. It is an extraordinary thing that, while Jacobite loyalty has found its bards, this loyalty to a thing far more closely linked with Highland history than a Lowland family ever could become, has not yet been sung.

In 1814, we have noted, all distillation from stills of less that 500 gallons capacity was prohibited within the Highland Line; the result of this piece of inequitable legislation was that in 1823 there were 14,000 charges of illicit distilling as compared with a total of six charges fifty years later. Still licenses were abolished in 1817 so far as Great Britain was concerned, but they remained in Ireland. The development of whisky distilling during the nineteenth century took very different paths in Scotland and in Ireland. In the former country the small, illicit home still was supplanted by large, commercial distilleries; in the latter, licensed distilleries grew fewer but the amount of illicit 'poteen' distilling probably remained constant.[18] In 1799 there were 87 distilleries in

---

[18] *There was a great flurry of pamphleteering by interested parties and other commentators in the late eighteenth century; this led to a major Parliamentary Enquiry which in 1798/99 issued two voluminous reports of 485 pages in total, with extensive illustrations and a map.*

Under the chairmanship of the indefatigable Rt Hon. Sylvester Douglas, evidence was taken, witnesses called and examined, and excise reports pored over, but the resulting legislation was largely ineffectual and overtaken by the 1822 and 1823 Acts. However, the 1798/99 Report is

Scotland; in 1817 there were 108; in 1825, there were 329; in 1833, there were 243; and in 1908, there were 150. In 1779 there were no fewer than 1,152 licensed stills in Ireland; by 1800 this had fallen to 124; and by 1908 to 27.[19]

So far as Scottish whisky is concerned, the era of the illicit distillers, the so-called 'smugglers', is of capital importance. For it was they and not the large-scale purveyors of cheap whisky who kept alive an educated taste in whisky and the true traditions of its manufacture. It is significant that the chief centres of the 'smuggling' have become classic districts of whisky, Glenlivet, Strathden,[20] the Glen of New Mill. Glenlivet became the centre of the smuggling industry of the North. The farmers of the glen, discovering that a demand for their product was springing up

---

an important record and demands a detailed appraisal in its own right.
[19] *There were draconian penalties for illicit distilling of poteen in Ireland during the early part of the nineteenth century. Despite a spirited defence of the government's policy by Aeneas Coffey, then Acting Inspector-General of Excise for Ireland (later the inventor of an improved continuous still) these were heavily criticised by the Rev. Edward Chichester in two powerful pamphlets. His* Oppressions and Cruelties of Irish Revenue Officers *brought to public attention the abuses which were further exposed in the important* Edinburgh Review *(March, 1819).*
Later, the Irish industry, once the largest in the world, faded dramatically. It is only in recent years that the Irish whiskey industry has been able to stage any kind of a meaningful recovery and once again take its rightful place on the world stage.
[20] *Strathden is presumably Strathdon, and Glen of Newmill is the Keith region.*

in the outside world, made up parties with trains of pack-animals laden with the products of their stills to carry the whisky over the mountains to the South. 'We have seen,' says a contemporary, 'congregations of daring spirits, in bands of from ten to twenty men, with as many horses, with two ankers of whisky on the back of each horse, wending their way, singing in joyous chorus, along the banks of the Aven.'[21]

To those pious law-breakers, who were more exercised in spirit because the necessities of their mystery compelled them to work on the Sabbath than because it brought them under the notice of the excise department, the modern connoisseur of whisky owes a deep debt. As late as the 'twenties of last century there was no legal distillery in Glenlivet, where 200 illicit stills were at work. The Duke of Gordon, who was the principal landowner in the glen, told the House of Lords bluntly that the Highlanders were born distillers; whisky was their beverage from time immemorial; they would have it and would sell it too, when tempted by high duties. But if the legislature would pass an act making it possible to

---

[21] *Quoted in* Lectures on the Mountains or the Highlands and Islands as they were and as they are *by William Grant Stewart (1860).*

The 'Aven' is the River Avon which runs from the loch of the same name to the Spey. Shortly before it joins the Spey is found the single-estate Ballindalloch Distillery, opened in 2014.

manufacture whisky as good as the 'smuggled' product on payment of a reasonable duty, he and his brother-proprietors of the Highlands would use their best endeavours to put down smuggling and to encourage legal distilling.

The government took the hint and the result was an act of 1823 sanctioning legal distilling at a duty of 2s. 3d. per gallon of proof spirit, with a license of £10 for each still over forty gallons. Smaller stills were made illegal.[22] This act is of capital importance in the history of whisky. It imposed moderate instead of grotesque burdens on the industry. And, at the same time, it killed whisky-distilling in the home as distinct from the factory.[23] But it could not have succeeded in this, had it not tempted enterprising

---

22 *MacDonald acknowledges the huge significance of the 1823 Act. He perhaps should also have mentioned the 1822 Illicit Distillation (Scotland) Act which set the scene for the more enlightened legislation which was to be passed in the following year. Also in 1823, the now familiar spirit safe, recently invented by one Septimus Fox, was made mandatory for licensed distilleries.*

Nor should Sir Walter Scott's part be forgotten. Scott orchestrated and stage-managed King George IV's state visit to Scotland in August 1822, an extravaganza of tartan and the conspicuous consumption of whisky that a grateful industry built on for the next hundred years and more.

23 *Perhaps not. Ambrose Cooper's* The Complete Distiller, *first published in 1757, was still in print in 1826, retitled as* The Complete Domestic Distiller *and described as for the use of 'distillers and private families'. And, as late as 1903, Mrs Charles Roundell and Harry Roberts would publish* The Still-Room *with its instructions for the domestic-scale distillation of 'Waters and Cordials'.*

individuals to undertake the manufacture of high-quality whisky. This, it was clear, could only be done in the areas where, as the 'smugglers' had proved, natural conditions favoured production. The first man to open a licensed distillery upon this sacred soil was Mr. George Smith, whose courage, both physical and commercial, should be remembered gratefully by all who have tasted that great whisky, 'Smith's Glenlivet'.

George Smith was born in 1792, the son of a Glenlivet farmer. He was an educated man of good family and the possessor of an excellent Latin style; he was in the beginning an architect and builder. But on his father's death he took over the farming of Upper Drumin, which, like his neighbours, he combined with illicit distilling of whisky. In the year after Waterloo the output of his 'bothy' was a hogshead a week. Mr. Smith has himself told the rest of his story:

'At length, in 1824, I, George Smith, who was then a powerful robust young fellow, and not given to be easily "fleggit," determined to chance it. I was already a tenant of the Duke, and received every encouragement in my undertaking from his Grace himself, and his factor, Mr. Skinner. The lookout was

---

In recent years, judging by the easy availability of small distilling plant from various websites, private distilling flourishes still.

an ugly one, though. I was warned before I began by my civil neighbours that they meant to burn the new distillery to the ground, and me in the heart of it. The laird of Aberlour presented me with a pair of hair-trigger pistols, worth ten guineas, and they were never out of my belt for years. I got together three or four fellows for servants, armed them with pistols, and let it be known everywhere that I would fight for my place till the last shot. I had a pretty good character as a man of my word, and through watching, by turn, every night for years, we contrived to save the distillery from the fate so freely predicted for it.

'But I often, both at kirk and at market, had rough times among the glen people; and if it had not been for the laird of Aberlour's pistols, I do not think I should have been telling you this story now. In 1825 and '26 three more small legal distilleries were commenced in the Glen; but the smugglers succeeded very soon in frightening away their occupants, none of whom ventured to hang on a single year in the face of the threats uttered so freely against them. In 1825 a distillery which had just been started at the head of Aberdeenshire, near the Banks o' Dee,[24] was burned to the ground with all its out-buildings and appliances, and the distiller had a very narrow escape

---

24 *This was the Banks o' Dee distillery which was indeed burned down by*

from being roasted in his own kiln. The country was in a desperately lawless state at this time. The riding officers of the revenue were the mere sport of the smugglers, and nothing was more common than for them to be shown a still at work, and then coolly defied to make a seizure.'

The task of the revenue officers in the Highlands, as they complained in 1784, was made more difficult because the distilleries were strategically situated on high ground which could not be overlooked. The surrounding lands were usually owned by the proprietors of the distillery who declined to allow the officers to build their houses in the vicinity. And, finally, large and fierce dogs were loosed upon any unwelcome visitors to the distillery.

The illicit distilling of Scotch whisky was not confined to the Highlands. There were 400 unlicensed stills in Edinburgh in 1777, when the number of licensed stills was only eight.[25] In 1815 a 'private' distillery of considerable size was found under an arch of the South Bridge in the Scottish

---

angry smugglers. According to Craig in The Scotch Whisky Industry Record *the precise location is today unknown. The danger continued for some time: in 1841, nearly twenty years after the liberalising 1823 Act, rival bootleggers burned down the original Lochnagar distillery after a former colleague decided to mend his ways and invest £10 in a distilling licence.*

[25] *Though frequently repeated, this figure must be considered suspect. It is derived from Hugo Arnot's 1779* History of Edinburgh *in which he writes that there are 'no fewer than four hundred private stills which pay no duty' in the city. However, ignored by MacDonald and all subsequent*

capital.[26] The only entrance was by a doorway situated at the back of a fireplace in the bedroom of a house adjoining the arch. Water was obtained from one of the mains of the Edinburgh Water Company which passed overhead, and the smoke and waste were got rid of by making an opening in the chimney of an adjoining house and connecting to it a pipe from the distillery. In the mid-nineteenth century a yet more scandalous instance of the Scottish aversion from paying duties occurred in Edinburgh when a secret distillery was found in the cellars under the Free Tron Church. But by that time the battle for legalised distilling had been won. It would no longer be the case that the finest whiskies could only be obtained by dubious and subterranean means from sources which could be but vaguely conjectured.

commentators, he then goes on to frankly admit that this estimate is 'only conjecture' and as Arnot was a fervent opponent of whisky drinking and wished to prohibit all private distilling his claim has to be treated with some caution.

26 *This wonderful tale appears in David Bremner's* The Industries of Scotland: Their Rise, Progress and Present Condition *(1869) which had first appeared as a series of articles in the* Scotsman *newspaper.*

The chapter on distilling appears to have been an important source for MacDonald. Remember also that around this time, under his own name of George Malcolm Thomson, he was researching and writing his major polemics on the condition of Scotland. With his keen interest in economic affairs and his robustly outspoken style MacDonald/Thomson will have recognised Bremner's earlier work as a useful basis for comparison between mid-Victorian industrial ascendancy and the distressed condition he observed in the 1920s and 1930s.

The modern history of whisky is so intimately associated with the development of manufacturing processes, with the commercial and financial aspects of what has become almost as much a science as an art and an industry rather than either, that it may be considered more conveniently under separate headings. We leave the story of whisky at a moment when it is exchanging a past illustrious and obscure for a present infinitely more prosaic, conducted in the full glare of modern commercialism and with all the devices at the disposal of a highly-capitalized, well-organized, large-scale industry. Whisky emerges from the shadows of the hermetic arts into the harsh limelight of the age of trusts and cartels and mass-production. The blue smoke rising warily above the heather dissolves and in its place there rises the gigantic image of one whose monocle, scarlet coat, top boots and curly-brimmed tall hat seem strangely remote from the glens and the clachans.[27]

---

[27] *The reference, of course, is to the iconic striding man of the Johnnie Walker brand, then and now the best-selling Scotch whisky in the world.*

The figure is believed to have been modelled originally from an old photograph of the real Johnnie Walker by the artist Tom Brown. His style has been followed in this advertisement by Leo Cheney.

Throughout the book, but particularly in this chapter, MacDonald would appear to have been heavily influenced by the work of J.A. Nettleton, in particular his pointed comments on the 'What is Whisky?' question and the work and findings of the Royal Commission.

Nettleton was an important figure—perhaps the most important—in the Victorian and early Edwardian distilling industry. His lasting legacy is his book The Manufacture of Whisky and Plain Spirit *(1913) which remained for many years the standard text on the subject and is still consulted today.*

He was utterly convinced that the Royal Commission had made a huge mistake in its recommendations and was a vehement critic of their verdict, describing it as 'a concession to a powerful commercial syndicate interested in the promotion of a modern innovation [which] sacrifices the interests of those who had obtained a reputation built up during centuries of development and enterprise to the interests of others who cannot boast of as many decades'. By this, he meant the DCL (the Distillers Company Ltd, forerunner of today's Diageo) who had actively promoted the cause of grain. It is strange that MacDonald does not specifically cite Nettleton's work but a close examination of the relevant texts leaves me in no doubt that he was heavily in the latter's debt.

Likewise George Saintsbury, revered amongst drinks writers for his Notes on a Cellar-Book *and one of MacDonald's professors at Edinburgh University, regarded the Commission as an exercise in futility and 'mischievous'. When we consider that it examined 116 witnesses over 37 days to produce a 724-page report that changed next to nothing we might concur with that verdict—other than the knowledge (with hindsight) that it was to profoundly influence the future shape and direction of the industry.*

The debate, of course, centres round whether that was for good or ill, and there is no doubt in which camp we would find Nettleton, Saintsbury and their eager student MacDonald.

III

MAKING AND BLENDING

Something of a noble and elegant simplicity characterizes the apparatus employed in the manufacture of whisky. Reduced to its bare essentials it consists of two main instruments, the *still* and the *worm condenser*. The still is a retort of copper with a broad, rounded bottom and a tapering neck. The worm condenser, connected to the still by a short pipe, is a spiral tube, also made of copper. There is nothing vital in the equipment of a distillery which a village blacksmith could not make and a small cottager could not buy. Moreover, no marked improvement of any note has been made in the machinery of whisky-distillation during long centuries. We can now make purer industrial alcohol, we have perfected the technique of treating the barley and making the malt, we understand something more of the chemical processes which take place when whisky is produced, we leave less to chance than our fathers did, but we have not devised a still which will make better whisky than the old pot-still which has been in use since the dawn of our knowledge of whisky, and we have not been able to construct a more

A promotional card for pure malt whisky (note age) from Sir Walter Gilbey's Glen Spey distillery. Gilbey wrote passionately in defence of the pot still.

practical method of condensing the vapour of the still [1] than the worm pipe kept cool by running water. The Highlander of the fifteenth or sixteenth century

---

[1] *The modern shell and tube condenser, first seen at MacDuff distillery c. 1962/63 is, of course, considered by its adherents to be a more efficient apparatus, but the worm tub continues to have its devotees, especially among the cognoscenti of single whiskies. Die-hard traditionalists are convinced of its superiority.*

Interestingly, and in support of that view, there have been instances (for example at Dalwhinnie distillery) where worm tubs have been replaced by shell and tube condensers but subsequently reinstated as the spirit character was felt to have changed unacceptably.

However, MacDonald is not strictly correct as there had been experiments at Hazelburn (1837) and later at Nevis, which will feature later in these notes, with partial condensers located on the stills or lyne arms. The idea that Scotch whisky production was unchanging until some unthinking commercial vandalism destroyed the values of a sacred priesthood is an appealingly romantic one, but not one that is supported by the historical record.

might be bewildered for a few moments if he were suddenly introduced into a modern malt distillery but he would speedily recognise in their larger bulk and greater elaboration the familiar instruments from which he was accustomed to obtain his beloved usquebaugh.

There are two main descriptions of whisky, depending on the raw materials used. Whisky is either made from barley malt alone (the great Scottish whiskies all belong to this class) or from a mixture of barley malt with unmalted grains of different sorts (all—or almost all—Irish whiskies come under this category). The process of distillation may best be studied by observing the manufacture of a pure malt whisky, such a whisky as will be found by the fortunate possessor of one or other of the classic brands. The essentials of the manufacture are as follows: [2]

---

[2] *Though the essential principles of distillation have remained unchanged since the earliest experiments, MacDonald gives here a description of practices in the 1920s. Since then, much advanced technology has been introduced, including highly computerised process controls and mechanical handling systems.*

Readers interested in a detailed description of the production techniques current when MacDonald was writing should refer to Nettleton; there are a number of contemporary texts describing the operation of distilleries in the twenty-first century.

Notwithstanding the introduction of much technology and despite the impressive scale of many of today's distilleries, I would hold, with MacDonald, that a distiller from centuries ago transported to a Roseisle or Inchbairnie would soon find the surroundings familiar.

The barley is brought from the farm into the barley-receiving room of the distillery where it is cleansed by being passed through screens; the smaller, inferior grains are discarded. The selected barley is then taken in bags into a barn from which it passes into the malt house.

Here it falls into tanks called 'steeps' where it is soaked in water. The softened, swollen grain is spread out on a malting floor[3] for about three weeks, being sprinkled with water at regular intervals and occasionally turned over. As a result of this, the barley begins to grow or germinate.

When the proper time has arrived the water supply is shut off; growth stops immediately and the grown barley is withered. In this state it is known as 'green malt'. What has actually been happening is that the starch of the barley has been partly converted into sugar by means of the ferment whose technical name is *diastase*.

The next step in the process is that of drying in

---

[3] *Few distilleries today maintain their own floor maltings; notable exceptions to this rule being Springbank, Bowmore, Laphroaig, Balvenie, Highland Park and Kilchoman. Even for most of these, a majority of the malt used will come from external suppliers operating large drum malting machines. The observant reader will have noted that four of the above have an island location—and Springbank is both small and remote from most major centres.*

Some of the latest generation of farmhouse or boutique distilleries have announced their intention to operate floor maltings.

the kiln. This is a supremely important operation for it is at this time that the malt acquires characteristics of flavour which it will later on impart to the whisky. Thus the chief distinction between the two main classes of Scotch whisky, Highland and Lowland, is that the former has its malt dried by means of peat fires. The combustion gases of the peat endow the whisky with that subtle 'smokiness' which is present, in discreet combination with other flavours, in genuine Highland whisky.[4]

The kilned malt passes into the mill room where a wire screen removes the 'culm' or sproutings produced during malting. The malt is ground in a mill. It is now passed into the mashing tuns where it is thoroughly mixed with warm water. Being later cooled, the resulting liquid, known as *wort*, is drawn off. The mashing process enables the diastase to convert the remainder of the starch in the malt to sugar and to *dextrine*, which dissolves in the wort. The wort passes through a refrigerator in the tun (or fermentation) room on its way to the wash backs, yeast being added to the liquor on the way. Fermentation now takes place, the sugary substance in the wort being transformed by the yeast into

---

[4] *This represents a marked change from today's practice where, with a few exceptions, peat (and 'smokiness') is not widely seen as a flavour note in other than island whiskies. It would not commonly be associated with Highland whisky in the modern era.*

alcohol and carbonic acid gas. The diastase converts the dextrine into sugar, which the yeast in its turn converts into alcohol.

The wash, or alcohol-containing liquor from the fermentation room, is now introduced into the wash still. This is the crucial operation, for now the liquid, leaving the world of earthly things behind, enters the realm of spirit. The result of the first distillation is a weakly alcoholic distillate known as *low wines*, which, after being condensed in the worm, passes to a receiver in the spirit-receiving room and is then returned to the spirit still in the still house to be re-distilled.[5] This time the distillate is collected in two parts.

In the spirit-receiving room is a glass case known as the spirit safe where the specific gravity of the

---

[5] *Though steam-jacketed stills had been employed from the 1870s, the stills MacDonald describes would have been heated by a direct flame, fired by peat, wood and coal (or a mixture thereof) according to local practice and the cost and availability of various fuels.*

Direct firing was so commonplace that MacDonald does not think to mention it; it is quite simply a given and thus entirely unremarkable. Today direct firing, usually by gas, is the exception to the rule. Also missing is any description of the rummagers in the wash stills: the chain-metal device that slowly scoured the inside of the lower part of the still preventing solids in the wash sticking to the copper and giving a burnt or empyreumatic flavour—though some commentators considered a hint of this to be desirable.

There is no mention of the use of soap in the wash stills to prevent foaming of the liquid which would then carry over into the lyne arm. Perhaps MacDonald was unaware of the practice, or possibly it offended his sense of the romantic.

distillate is tested. The first portion of the second distillate is whisky and is passed into the spirit receiver; the second, known as *feints*, contains alcohol in considerable proportion and passes into the low wines receiver where it is mixed with the next charge of low wines for re-distillation.[6]

The finished whisky flows into a spirit vat in the spirit store and is poured into wine-saturated casks where it will mature. The function of these casks—they are generally sherry casks—is often misunderstood. They do not impart qualities to the whisky which it did not possess before, although the action of a saturated cask may be to conceal, under a strong, imported, winey flavour, some defects in an inferior whisky. But the casks give colour to whisky, which in its native state is a wholly colourless liquid.[7] Why this austere achromatism of whisky should be

---

[6] *This is really quite a cursory description of distillation, here omitting mention of the foreshots and heads before the spirit cut—that part eventually destined to become whisky—is taken.*

[7] *What to say? This is plainly wrong. In pleading for MacDonald he is evidently an amateur and it is the case that the understanding and appreciation of the role and contribution of wood has grown immeasurably in recent decades. He is, at least, correct about their role in adding colour. Best to move on ... But not perhaps until noting that he maintains the wood then in use to be 'generally sherry casks'. A very great quantity of sherry was transported in cask to the UK at this time, a trade which has now ceased due to the decline in demand for sherry and legislative changes. Today a significantly greater number of ex-Bourbon and refill whisky casks are used; small numbers of casks formerly used for port and various wines are also used in finishing.*

unpopular is not easily to be explained. It may be that the unearthly pallor of the pure and fiery spirit strikes terror into the heart of man, as the whiteness of the whale alarmed Captain Ahab and his ship's company. (But why, then, does no one protest against the colourlessness of gin?) It may be that whisky was originally a 'doctored' spirit, containing in addition to malt distillate, spices and other extraneous flavouring and colouring matter. It may be that the idea of drinking whisky made solely from malt or grain is quite modern. There is, indeed, some evidence for a popular, as distinct from a learned, taste for spiced usquebaugh.

In George Smith's *A Compleat Body of Distilling*, published early in the eighteenth century, there is a recipe for 'Fine Usquebaugh'.[8] It contains six gallons of proof molasses-spirit, six gallons of proof rectify'd spirit, and a whole gorgeous East of condiments—mace, cloves, nuts, cinnamon, ginger, coriander seed, cubebs, raisins, dates, liquorice, best English saffron, and Lisbon sugar. Clearly the ingenious Mr. Smith was only providing a home-made substitute for usquebaugh, but he could hardly have included spices unless he were seeking to imitate flavours which could hardly exist in an unadulterated

---

8 *The term 'usquebaugh' seems to have been employed with less discrimination in Smith's time.*

whisky. His inclusion of saffron is also significant, for this would give a colour resembling that of modern whisky. In the North and the West, where spices were hard to come by, pure whisky, deriving its flavour from moss-water, peat and barley malt and from nothing else, first came into being and its high qualities were first recognized. But the prejudices of the Lowland and English market had to be considered, and so (we may conjecture) dyed whisky was manufactured. To this day Highlanders do not exhibit the aversion from colourless whisky that is so common among Lowlanders and Englishmen. (And, it may be observed in passing, fluid caramel or paxarette is frequently added to give whisky the complexion which it is supposed to derive from residence in sherry-casks.)[9]

It is not to be thought, however, that the wine-casks do nothing more to the whisky than give it a tint. The wine-soaked wood of the casks absorbs various insoluble bodies which would

---

[9] *Much of the preceding may be considered a 'conjecture' too far, though in* Whisky and Scotland *(1935) Neil Gunn refers to the consumption of 'clearic'.*

Paxarette (effectively a sherry concentrate used to prepare casks prior to filling) was prohibited by the 1990 Scotch Whisky Orders but had not been widely used for some years prior to that. Spirit caramel E150a is still a permitted additive for colouring purposes but remains controversial among certain groups of consumers and a number of brands make a point of emphasising that they do not use it.

Cameronbridge Distillery was established by John Haig in 1824. Today, operated by Diageo, it is the largest operation of its kind in Europe

otherwise mar the flavour of the matured whisky.[10]

The manufacture of Irish whisky closely resembles that of Scotch except that, instead of barley malt, a mixture of malted and unmalted barley, wheat, oats, or rye, is mashed together. The diastase in the malt is sufficient to cause the breaking down of the starch in the unmalted grain into sugars, the subsequent conversion into alcohol being performed by yeast,

---

10 *The reference to the extractive properties of casks would seem to contradict MacDonald's earlier assertion that casks were largely inert, though he would doubtless argue that removing a fault does not impart a quality, merely removes an imperfection.*
Rather more is known today about the action of wood, and detailed scientific studies continue in the R&D departments and laboratories of all the major distillers and at the Scotch Whisky Research Institute near Edinburgh.

as before. The process of distilling is carried out in pot-stills, whereas the Scotch grain whiskies are made in patent-stills.

The first patent-still to come into wide use was that invented about 1826 by Robert Stein[11] of the celebrated Scottish family of whisky distillers. At that time there were 114 distilleries in Scotland, of which five were making for the English market. All of these were owned by the Stein family or by three firms of the name of Haig. Stein speeded up the working of the still by bringing steam into contact with the wort. The idea was developed later in the famous patent-still of Mr. Aeneas Coffey of the Dock Distillery, Dublin. This was invented in 1830 and replaced Stein's patent. Coffey's still consists of two columns, the *analyser* and the *rectifier*. In the first of these, previously heated wort is passed downwards over a series of perforated copper plates, and meets an upward current of steam which carries off the alcohol from the wort. This is condensed at the bottom of the rectifier (the second column)

---

[11] *The same Stein of Kilbagie of whom MacDonald is so dismissive earlier in the book. Stein was not alone in experimenting with the continuous still. Adam, Fournier, Blumenthal and others all developed similar apparatus.*

Eventually, the Coffey design became the most successful and widely used in Scotland, though Stein stills were operated until at least 1860 before being supplanted by the Coffey design. According to Nettleton, 'the ruins [of Stein stills] were standing at more than one distillery as late as 1905, and may be yet'. He was writing in 1913.

on another series of perforated plates, this time cool plates, where it is collected in two portions, an upper consisting of more volatile alcohol and a lower known as *hot feints*. By this type of still it is possible to produce in one continuous operation a spirit containing 95 per cent. of alcohol. Coffey's still is therefore the means by which the greater part of our industrial alcohol is manufactured. Its value in whisky distilling is more doubtful. For whisky derives its character—or, rather, whiskies derive their characters—from the presence in the finished liquid of tiny 'impurities' which the hard efficiency of the patent-still eliminates.

The problem of whether or not spirits made in a patent-still were entitled to the name of whisky is one which at one time exercised the minds of those who were genuinely interested in the honour of the liquid.[12] It is difficult not to regret that the battle went on that occasion to the strong in numbers and wealth, and that the final decision was in the hands

---

12 *The reference is to the 'What is Whisky?' controversy of 1905/06 and the subsequent Royal Commission on Whiskey and Other Potable Spirits which followed. Its membership was largely comprised of English medical experts, leading to some complaints from Irish MPs. The trade were deliberately excluded from membership to avoid the charge of bias, but called to give evidence before the commissioners and did so in considerable numbers.*

The detail of the Royal Commission and the contrasting stances adopted by the grain distillers (essentially the Distillers Company Ltd) and representatives of the pot still industry is too complicated to relate here but, as

of those who were ignorant of what whisky really is.

It is unfortunately true that the whole modern history of whisky is a record of the opening of door after door to commercial vandalism, of the stretching of definitions until they cease almost to have any meaning at all.[13] The influence of the patent-still is not the only one to be deplored. Whisky has shown, or has been compelled to show, such a hospitality to strangers that its house has now an excessively crowded and variegated appearance. As early as 1678 the principal materials used in making Scotch whisky had been fixed, though herbs and spices were probably added as in Ireland to produce the various local brands of aqua vitae, cordial, etc. Sir Robert Moray writes,[14] 'Malt is there (in Scotland) made of no other grain but barley whereof there are two kinds, one which hath four rows of grain

---

MacDonald makes clear, the result was to promote the interests of blending.

It is interesting to note that, whatever the shortcomings of the enquiry, it did at least establish a recognised legal definition of whisky, even if one that not everyone agreed with or felt accurate or fair. In doing so, the Royal Commission laid the foundation of the modern industry, and it is an irony that even MacDonald would appreciate that the survival of so many single malt distilleries today owes much to the success of the blends (and brands) that he so vehemently deplores.

[13] *One might argue that the industry's survival and commercial success owes much to the influence of the patent still and blending—not that our author appears inclined to reasoned argument. This is passionate advocacy at its most intense.*

[14] An Account of the Manner of Making Malt in Scotland *(1677)*.

in the ear, the other, two rows. The first is more commonly used; but the other makes the best malt.' In the first half of the eighteenth century the 'Society of Improvers in Agriculture'[15] took up the question of whisky distilling; one James Dunbar[16] made suggestions about the manufacturing process and a Dutchman, Wyngaarden, drew up directions for the use of Scottish distillers. Unscrupulous manufacturing methods were not a crime of the Highlanders, who protested that the spirit made in the Lowlands, and in Ireland, from unmalted grain was 'Scotch'd spirit', not whisky. Until quite late in the century Irish usquebaugh was a cordial made by adding cinnamon, liquorice, and other spices to grain or malt spirit. It seems possible that in Scotland aqua vitae referred to a pure malt spirit, and usquebaugh to a spiced cordial; at any rate a distinction of some sort was preserved, for in 1732 the Duke of Atholl wrote to Lord George Murray, the famous general of the '45, complaining that 'I have not one drop of either usquba or acquvitae in the house.' In the early part of the nineteenth

---

15 *This would appear to be the Select Society for the Encouragement of Arts, Sciences, Manufacture and Agriculture, who in 1755 offered a 'premium' for the best tun of whisky.*

16 *Though he may have been a more prolific author, I can only trace Dunbar's work* Smegmatalogia, *or the manufacture of soap. De Wijngaardenier was a Dutch vintner and brandy distiller operating in Amsterdam, c. 1736. He appears with his still in a copper engraving by Jan and Caspar Luyten.*

WHAT IS WHISKY?

ASK A TEMPERANCE REFORMER.

One of a set of six contemporary postcards satirising the work of the experts of Royal Commission on Whiskey. Thomson's mother was a staunch temperance advocate but, more than one hundred years on, the image bears a striking resemblance to a well-known whisky writer!

century the words 'Scotch whisky' had a very definite and precise meaning.[17] They denoted a whisky made in Scotland from malted home-grown barley (dried over a peat fire in the case of Highland whisky), and distilled in a pot-still. At the same time Irish whisky had a similarly close and exact definition. It meant whisky made in Ireland by treble-distilling, in pot-stills heated by furnaces, of a wort obtained from a mash of malted barley, barley and oats, all the grain being home-grown.

As the century progressed, however, it became obvious that these definitions could not be adhered to. By the '70s it could no longer be concealed that much of the barley used in making Scotch whisky was grown abroad. This was a grave breach with tradition, for as early as 1703 the bailies of Strath Spey decreed that aqua vitae was to be brewed only from malt grown in the locality.

The introduction of patent-still distilling had an even more disturbing effect in the art of whisky-making. For if a patent-still spirit was a whisky, then where was the line to be drawn? Only by

---

17 *This may have been the case through custom and practice and the de facto case in the absence of any alternative, but there was no such legal definition while there was a problem of unknown extent with counterfeiting and adulteration at a wholesale and retail level. However, as late as 1888 the ninth edition of the* Encyclopaedia Britannica *states that 'malt whisky is the product of malted barley alone, distilled in the ordinary pot-still'.*

wresting the name 'whisky' violently from its earlier, well-established meaning could it be applied to the neutral, 'silent', flavourless product of the patent-still. Adam Young in his *Distillery Instructions*[18] refuses the name whisky to patent-still spirits; as late as 1890 it was widely and authoritatively denied.

But the Royal Commission on Whiskey of 1908–1909 opened the dykes to the invading floods of patent-still spirit by defining whisky as 'a spirit obtained by distillation from a mash of cereal grain, saccarified by the diastase of malt.' This was comprehensive enough, and sweeping enough. It made no distinction between Scotch and Irish whiskies; it did not stipulate that the grain or malt should be home grown; it did not specify that pot-stills alone were to be used; it did not mention Scotch whisky's true character as a distillate of a wort made

---

18 *This is a rather obscure technical work, issued anonymously in 1880 but under the general supervision of Adam Young, then Deputy Chairman of the Inland Revenue. He retired in 1886, dying in 1897. Nettleton praises him lavishly, considering him 'the greatest man the Excise service ever produced'.*

The book—correctly titled Instructions on Surveying Distilleries and Charging the Duty on British Spirits—*is not one with which I would have expected MacDonald to be familiar, but Nettleton expounds upon it, making (at length) exactly the point that MacDonald summarises so trenchantly here.*

The significance of the date 1890 is that it marks the earlier Parliamentary Select Committee of 1890–91 that first attempted a definition of whisky and accepted grain spirit as within that term.

from malt. As one commentator[19] has sardonically observed, it was at least something that beetroot, potatoes, and sawdust were eliminated. They may have been by the sagacious Commission but at least one of them, potatoes, is mentioned in an exhibit at the Science Museum, South Kensington, as a source from which whisky can be obtained!

The reckless extension of the term 'whisky' has had the gravest consequences for the prestige of the industry. It has tended to deprive whisky of the special character it had built up during centuries of careful and pious labour and research. The tasteless distillate of grain, made at one process in a patent-still, is equally entitled to call itself whisky as the exquisite, pot-still, malt whisky, dried above a peat fire. It is only right to say that the definition was made in defiance of the best opinion of the distilling industry. The Duke of Richmond and Gordon, landlord of the most famous of all whisky districts, gave expression to this opinion in a speech in 1909: 'Quite recently, a public inquiry has taken upon itself to decide, What is Whisky? And I regret to say that apparently anything made in Scotland, whatever its combination, is to be called Scotch whisky. But for my

---

19 *The commentator was Nettleton. Like much of this chapter, this and the following paragraph are derived almost directly from his* Manufacture of Whisky.

part, I should prefer, and I think most of those whom I am addressing now would prefer, to trust their own palates rather than to the dogma of chemists.'

At the same time, it is to be deplored that the opposition of serious-minded, cultivated whisky drinkers to the pretensions of the Whisky Commission was not stronger. It was nothing short of a sin against the light to lump malt whisky with neutral industrial spirit as if it too were something to burn in lamps, to drive engines, or to clean clothes. The evil having been done, however, it is necessary to instruct the whisky public, especially young and inexperienced drinkers, in the true facts of the case so that, so far as possible, 'whisky by grace of the Royal Commission' may be left to those who ask for nothing more from their beverage than a 'kick'. This, at least, it will guarantee for them. But the children of light will continue to demand of their Scotch whisky that it should be distilled in Scotland by means of pot-stills, from mashing materials consisting of malted barley and nothing else, dried in kilns by peat or other fuel according to the locality; and of their Irish whiskey that it should be pot-still, from malted barley, either alone or with unmalted barley, oats, rye, or other indigenous cereal. By accommodating themselves thus far to modern conditions, they will assure themselves of a whisky which it will be possible to drink without a grimace (with a heroic determination to overlook the means for the sake of the end), and even—granted discrimination—with delight.

The distilling of whisky is only one half of the manufacturing process. We have still to account for the fact that though there are less than 130 Scotch and Irish distilleries, there are over 4,000 brands of whisky on the market. Whence comes this monstrous multiplication? The answer is to be found in the process known as blending.

The practice of blending whiskies sprang up in Scotland about 1860 as a natural development of a process which was much older, 'vatting'. By 'vatting' is meant the mixing of single whiskies from the same distillery but belonging to various distillations made at different periods of the year. Its purpose is to obtain a whisky of uniform character. For whisky distilled in October differs from a winter whisky and both of them differ from a spring whisky.[20] Owing to changes in temperature, in the character of the barley, and in the chemical composition of the water employed, whisky, that master of delicate

---

20 *Consideration of the impact on spirit character of different strains of barley is today a hotly debated topic. Some distillers, generally the larger firms making for substantial blended brands, hold that it is of little or no significance while others maintain with perhaps greater fervour that it is of considerable importance and that different flavours can be detected in the new make spirit when different varieties are employed.*

Both positions are defensible, for while it is demonstrably the case that

adjustments, differs from one week to another.

About 1853, spirits from the same distillery but of different ages were already being blended; by 1865 pot-still and patent-still (*i.e.* grain) whiskies were blended for the home market. The pioneers in the revolution—for it was nothing less—were Messrs. Andrew Usher & Co. of Edinburgh whose old-vatted Glenlivet achieved a considerable reputation about that time. The practice spread and soon Glasgow, Leith, Dundee, and Aberdeen were the chief centres of a considerable blending industry. From the purely commercial point of view, blending was a tremendous benefit for the distillers. The old single malt whiskies of the Highlands were, on the whole, too powerful and heavy for sedentary town-dwellers. Blending made it possible to make a whisky which would suit different climates and different classes of patrons. For by adding the lighter Lowland malts and the neutral or almost neutral grain whiskies, in greater or less degree, a whisky could be evolved of a 'weight' and a strength of flavour and bouquet

---

different varieties do give different spirit character (and this has been proved to my satisfaction in distilleries in both Islay and Ireland) it is also true to say that such subtleties are swallowed up in the blending process, the blender's aim being to produce a consistent product.

The topic is further complicated by the undeniable influence of the finance department who constantly urge improvements in yield to increase profitability. Hence the introduction of many of the new barley strains, a trend and an influence that MacDonald would doubtless deplore.

to suit the taste or the commercial requirements of the blender. The great export trade in whisky is almost entirely due to the adaptability and elasticity which blending lent to the industry. Even today the aesthetics of whisky have a very definite geographical aspect. London likes a milder, less pronounced whisky than Lancashire. Lancashire in its turn affects a whisky which is lighter and less pungent in taste than that which solaces the east winds of Edinburgh. But Edinburgh is surpassed by Glasgow, where they revel in the 'denser' and fuller-bodied joys of the Campbeltown malts. In the Highlands, malt whiskies are still drunk, uncontaminated by the diluting, chilling alliance with grain. In the country districts where the distilleries are to be found, there are devotees of the old 'single', 'self', or unblended whiskies. As for the export markets, Australia and India favour the lighter, drier charms of the blends *goût anglais*, in which the Highland malts have been subdued by grain and Lowland malt. But Canada, as one might have expected, tends to follow Scotland. On the continent, the Scandinavian countries display a preference for the class of whiskies which find favour in Edinburgh and its neighbourhood; but Holland, Belgium and France, where it has become smart of recent years to ask for '*le scotch*', take their opinions on this important matter from the English.

The technique of blending can be briefly described. The first style may be termed the 'rough and ready'. The proportions of the different whiskies to be used are determined beforehand, in most cases upon a tried formula; the whiskies are poured into a vat where they are mixed up thoroughly and allowed to stand for a day and a night in order that the 'marriage' may be consummated. The blend is drawn off into casks and bottled shortly after. The result is a cheap whisky, imperfectly blended, raw and unsatisfactory in flavour. The better practice is first to prepare blends of carefully selected malt whiskies and allow these to stand in casks for about two years. When this has been done, it is a comparatively simple matter to make a blend that will suit an individual market. The blended malts are run into the vat and mixed with more or less Lowland malt and more or less grain whisky. The resulting blend is then drawn off and left in the cask to marry for a minimum period of six months. The process of mixing in the vats, which used to be done by revolving paddles, is more usually performed in large-scale blending by the passing of compressed air through the mixed liquors from a large number of small jets in the bottom of the vat. In this manner, it is said by some, a surer and more intimate mixture is obtained and the horrid possibility of a subsequent failure of marriage banished. But no technical device affects the supreme clause in the matrimonial legislation of whisky, that time is of the essence of the contract. Whiskies are capricious, sensitive creatures; they are not to be flung at one another like goats. Rather are they to be compared to fillies which are highly likely to plant iron heels in the belly of the too-forward stallion. They must grow accustomed to one another and, unless they have been carefully chosen, no amount of time will persuade them to live together in amity.

After the experimental stage of blending had passed, the blender evolved fairly definite rules of his craft.[21] He no longer works in the dark; he knows exactly what he is going to create. He begins by grading whiskies in territorial groups. Thus—to anticipate somewhat the contents of a later chapter—the whiskies of the Glenlivet district and its surrounding country (for only one whisky is legally entitled to describe itself as 'Glenlivet', *nomen praeclarissimum*) are ranked together; the North Country whiskies (for the most part Aberdonian) form another class, and so on. Within certain limits the blender is aware that he can substitute

---

21 *This process of marrying is described by Alfred Barnard in a pamphlet from the 1890s. The practice of first pre-blending the components is often attributed to A.J. Cameron of John Dewar & Sons, around 1914.*

one North Country whisky for another in his blend, one Speyside for another, one Islay for another. He also knows that, although there are excellent malt whiskies which would never, were they to be kept to the Judgment Day, do anything but bicker in the cask, there are unions predestined to be blessed. Thus the Banffshire whiskies can be counted upon to make a happy and enduring marriage with the Islays.

Although blenders preserve jealously the secrets of the different ingredients and their proportions in the blends, it may be taken that the whiskies which are drunk over the greater part of the world are made up from equal quantities of Highland malt whiskies and grain whiskies, with the addition of a smaller proportion of Lowland malt. Islays or Campbeltowns are also used, but they should be previously blended with the Highland malts to which they will add their own noticeable but pleasant flavours. The object of the good blender is to obtain a smoothly finished liquor in which no individual strain will thrust its characteristic taste or aroma upon the palate. The unscrupulous blender, catering for the unthinking multitude, will cheaply and rapidly obtain a whisky of sorts by adding a small quantity of a strongly flavoured Islay or Highland to a large quantity of cheap grain whisky. But the good blender (may heaven be kind to him and grant him skill!)[22] knows that no whisky that an honest man could sell without blushing is to be got by such crude means. And here it is that the Lowland[23] malts play their part in the scheme of things. It is their function—for they are but slightly flavoured liquids—not to add an instrument to the well-balanced orchestra, which every fine blend is, but rather to act as the conductor, giving a steadiness of rhythm to the music. They are, as it were, catalysts; they act as a bridge between the pungent Highland malts and the sexless, neutral grain whiskies; without their presence a fine, smooth blend is impossible.

Blends of malt whiskies without the addition of grain are practically unknown so far as the ordinary public is concerned. About twenty-five years ago, when the Highland distilleries had large stocks in hand, an attempt was made to create a market for blended malts. The attempt failed, in spite of a vigorous advertising campaign. People who had acquired a taste for the more insipid charms of the grain-containing blends were not to be won by the

---

22 *Here, at least, MacDonald seems to extend some generosity to the blender.*
23 *Alfred Barnard, eulogising the merits of Islay whiskies (in a promotional brochure for his client Peter Mackie, featuring Laphroaig and Lagavulin distilleries), describes the best makes of Lowland malts as 'useful as padding'. They apparently 'help to keep down the price of a blend' and are 'decidedly preferable to using a large quantity of grain spirit'—damned with faint praise one feels.*

more definite flavours, the wider range of variations, the greater subtlety of the unblemished malts. The more's the pity. Perhaps we are not manly enough to grapple with the really great whiskies. But some there are who have not succumbed to the degeneracy of the age. There is, for example, the patriarchal Professor Saintsbury who records[24] that the best of all the blends he has tasted was one of Clyne Lish (a Sutherland whisky) and Smith's Glenlivet. Grain whisky the same authority very properly describes as fit only for drunkards and for blending.

Irish whiskies are not blended so frequently as Scotch. The best opinion is that the Irish distillations are better as 'self' or single whiskies.

The adulteration of whisky might well have a chapter to itself—if it were not for the fact that the consumer of a reputable whisky need not fear to be its victim.[25] Vitriol was at one time added to whisky

---

[24] *Strictly speaking, Saintsbury records the opinion of 'a friend of mine from Oxford days, now dead' on this blend as 'the best whisky he [i.e. the friend] had ever drunk'. He did, however, dismiss grain whisky in exactly the terms described. Presumably the heavily tattooed ex-footballer turned champion of a lavishly promoted brand of blended grain whisky is not a student of either Saintsbury or MacDonald.*

[25] *It may be that by 1930 MacDonald was able to take this sanguine view. However, it is undeniable that during his 'golden age' of whisky there was a problem with adulteration at both wholesale and retail level.*

The adulteration of whisky in the retail trade in Glasgow was thoroughly documented in the 1870s by the North British Daily Mail *(see Edwards Burns'* Bad Whisky, *Angel's Share, 2009). The Irish pot still distillers*

(in very small quantities, needless to say) in order to improve the 'bead' or bubble proof of the spirits. 'The doctor', by which was meant a mixture of vitriol and oil of almonds, was sometimes called in by Highlanders who wished to see a froth in their casks. The same effect was artificially produced in Canadian spirits by adding glycerine and 'bead oil'. When the United States still manufactured whisky (or at least a liquid whose local name was whisky) a pleasing concoction called 'Bourbon oil' was sometimes added to give it a frothy appearance. Bourbon oil consists of fusel oil, acetate of potassium, sulphuric acid, and other equally attractive ingredients. If such things could be done before Prohibition, one need not be surprised at anything that happens after.

---

railed against the indiscriminate use of 'Hamburg sherry', 'prune wine' and 'cocked hat spirit' in whiskies, and, as late as 1911, the Encyclopaedia Britannica *could write: 'A common form of adulteration of whisky is the addition to it of spirit made on the Continent mainly from potatoes.'*

The related problem of counterfeiting continues to plague the industry and defraud the consumer.

IV

GEOGRAPHY

It is fairly well known by this time (or some painstaking and persuasive authors have written in vain) that the wines of Europe are to be grouped in various more or less closely defined geographical areas. By degrees the assiduous oenophile learns to discriminate, in his infancy between Burgundy and Bordeaux, in his hobbledehoy days between, say, Mâcon and Beaujolais, and, when he reaches years of vinous discretion, he has educated his palate to the nicer expertise required to detect a Romanée Conti from a Richebourg. But who appreciates that there are geographical divisions in whisky? Too few, it is to be feared. Yet, even as there are in the kingdom of Bordeaux fair provinces, Médoc, Graves, Sauternes, Saint-Emilion, Pomerol, for this polity of the wine is no colourless cosmopolis but, indeed, a confederation or bund, so within the ethereal realms of whisky boundaries are fixed, marking off one lordship from another; there are territorial families in the north, at the fashion of whose marryings and giving in marriage we have already glanced. The geographical aspect of whisky is one which must receive the consideration of all who regard it as something more than a liquid whose taste, presumably offensive, one destroys with gaseous waters out of syphons.

One does not pretend that the whole explanation of whisky's divine variety is to be found in geography. In an operation so delicate and elaborate as distilling it is obvious that a considerable amount must be left to the highly-skilled technicians (artists one should rather call them) who control the mystery. But this would not account for the undoubted—though still somewhat puzzling—fact that whisky falls into well-defined classes corresponding to geographical areas.[1] Geography exerts an influence, secret and subtle, upon whisky—and so far no one has been able to determine through what precise media it operates or to what degree each of the supposed factors is to be held responsible.

A well-informed writer in a trade publication puts the position with admirable succinctness: 'The malt distilleries ... are surely among the most remarkable phenomena in British industry. Why a score of Highland distilleries should produce twenty slightly different types of whisky, all good and all apparently

---

[1] *While geographical classifications remain of some value it is broadly true to say that they are less important today than was the case before the war. This section is therefore of greater historical interest than current relevance, though the reader may find a discussion of lost distilleries, such as those of Campbeltown, more than a little poignant.*

inimitable, no one seems to know with any certainty.' Despairing of an adequate explanation of this along purely territorial lines, he arrives at the conclusion that the credit for the differences between one whisky and another is to be attributed to the nature of the utensils employed and to the local variations in the method of working, variations which, being preserved from generation to generation through the reverent conservatism of distillers, secure the permanence of each whisky's character. Thus, the shape of the still in use has an undeniable influence upon the whisky distilled from it, for upon its shape depends the proportion of essential oils and higher alcohols which will be collected in the condenser. And, on the other hand, the still-man, that almost priestly functionary, fills a rôle of the highest importance.[2] He has the final voice in determining the style of a whisky (within the limits imposed upon him by his utensils and his material) because so much depends on the precise instant, in the course of distilling,

---

[2] *With increased mechanisation and greater and tighter process control, it may be conceded that the consistency of spirit output has been greatly improved since MacDonald was writing. However, his romantic view of the still-man as the possessor of arcane knowledge acquired over years, decades even, of dedicated labour still appeals to the aficionado of single malt and is exploited in the marketing literature of more than one distillery. It is a powerful, appealing and evocative image—even if the reality of the still-man's daily routine is more closely related to the monitoring of a computer screen from behind the pages of a newspaper.*

when, acting upon an intuition securely fortified by experience and tradition, he stops collecting the finished spirit and returns the remainder of the distillate to the still. Custom in manufacture, inherited and jealously guarded, necessarily plays a considerable part in bestowing traits of one kind and another upon a liquid which at the crucial moment of its birth assumes the delicate and impressionable form of a vapour. It is even held by some old distillers that a finer whisky is made if peat is used *to heat the spirit-still.*[3] Far-fetched as it may appear, there is almost certainly something in this contention, for peat does not give such a high temperature as coal, and therefore the wash in the still is brought more gradually and steadily to the boil.

When everything has been admitted, however, in a question where the last word must remain with the chemist (always assuming that a chemist can be found capable of appreciating the more esoteric niceties of whisky's flavour and aroma, as well as of conducting experiments with test tubes and retorts) there remains the practical consideration that there are territorial classes of whisky. And it is not difficult to understand why there should be, even if the persistence of

---

[3] *Until a very traditionally-minded distiller, most probably from the recent generation of 'craft' producers, can be persuaded to experiment in this way it would seem that the taste of whisky distilled in a peat-fired spirit still must remain in the imagination.*

local distinctions in manufacture be discounted.

There are four main geographical factors brought to bear upon whisky.

First, there is air. The relative purity, humidity, etc. of the atmosphere in and around a distillery is not the least important of the character-bestowing agencies that affect whisky. Practically all the malt distilleries of Scotland are located outside large towns; the most renowned of them stand in bare, open country, far from the smoke and the dust. The great belts of wood and heath which are characteristic of the Banffshire and Morayshire uplands cleanse the winds which sweep across them and impart some of their own purity to the whiskies of hallowed name that are distilled there. Nor is it without significance that about fifty per cent. of the Scottish distilleries stand within a few miles of the sea. There seems to be some obscure association between the qualities of sea-air and the whisky produced in it, although there is no better evidence for the opinion than the fact that a large number of distilleries flourish almost within sight of salt water.

Second, water. This is so important that it scarcely needs to be mentioned. Yet the effect of water upon the style of a whisky is negative rather than positive. Its virtue consists in adding no flavour—or at least no marked flavour—to the malt which it makes. It should be pure, colourless, without odour, free from micro-organisms or grosser organic matter, above all devoid of mineral contamination, dissolved or in suspension. Each distillery recites the praises of its water-supply.[4] Highland Park, the well-known Orkney distillery, obtains its water from two hidden springs: 'it never sees the light from source to mash dam'. The Glen Elgin-Glenlivet distillery claims to possess (though, indeed, it is not alone in making the claim) the finest distilling water in Scotland, drawn from a mountain spring. The Glenlivet distillery relies upon 'the finest and purest water on earth, which tumbles down the mountainside for 1,200 feet, and glides through the district in the sparkling stream of Livet'. The Old Bushmills distillery, one of three Irish distilleries making a pure malt spirit, employs the water of the river Bush, which flows through an extensive peat bog. There seems to be no doubt that the effect of such water upon the malt will be somewhat similar to that of the peat which the Highland distiller burns in his kiln. But, as a rule, the distilling water should conform to the specifications set out above. The purer the water the better will it perform the task of converting starch into diastase,

---

[4] *Distilleries do indeed praise the quality and purity of their water supply, though it seems to me that we have heard less of this in recent years than once was the case. Wood, one might almost observe, is the new water.*

maltose, etc., and the purer will be the subsequent vinous fermentation.

The geographical bearings of all this are obvious. Percolating through subsoil of a certain uniform character, flowing into faults in rock of a specific geological nature, and, finally, running over the stones and earth of river-beds descending from the same watershed down the slopes of the same hills, thus does rain-water acquire a common body of local or provincial qualities. The best and the most productive whisky-district in the world is that which is served by waters flowing from the hillsides of the Cairngorms and their sister outposts of the Grampians in Banff and Moray. Let it not be assumed, however, that the same hillside will at all points yield the same water. There was once a distillery built at considerable expense in a West Highland seaport[5] to make use of the waters of a burn flowing from a most august mountainside. The water was all that

---

[5] *This would seem to be a reference to the Nevis distillery in Fort William, built in 1878 to meet the growing demand for whisky from the nearby Ben Nevis distillery, which was then in the same ownership.*

Barnard writes about both, but devotes more space to the Nevis distillery which, at the time of his visit, was the larger of the two, producing around seventy per cent more whisky than Ben Nevis. Unusually, the still-house in Nevis is illustrated in Barnard and shows a curious arrangement of partial condensers on the wash stills and what appears, at a tantalising distance, to be an experimental spirit still.

Paradoxically, while he does not specifically mention the Ben Nevis water

could be wished for. It was clear and sparkling to the eye, pleasant to the palate, triumphant in the laboratory. But alas! It had one fault. Good whisky could not be made from it. Chemists, maltsters, and still-men could try as they might: it was of no avail. The money, it seemed, had been as good as thrown away; the glittering deceitful Highland burn had charmed thousands of pounds out of investors' pockets, for nothing. But a mile away there was another burn, this time a mere trickle of water and not particularly tempting to look at. Despair suggested an experiment with its water, which came from the same slopes as that of the deceiving burn. The result was astonishing—a whisky of high quality. It is widely used to-day in blending, and the buildings of the original distillery are now a storehouse for the

---

source, Barnard goes to considerable length to praise the water used at Nevis, even quoting an analyst's report and noting that it is the same source that supplies Ben Nevis.

What are we to make of this? Was the proprietor anxious perhaps to scotch any lingering rumours about the quality of the original water? In any event, by 1908 it appears to have been closed and, as MacDonald relates, the buildings were in use as warehouses for Ben Nevis.

While one cannot conclude absolutely that this report relates to the Nevis operation, tales of poor water quality do on occasion attach themselves to various distilleries. A similar story circulated in connection with Glenglassaugh—without, as it happens, any basis in truth. An earlier version of this anecdote appears in Irish whiskey's literature in the 1925 pamphlet Elixir of Life *issued by John Jameson & Sons with illustrations by the noted Irish artist Harry Clarke.*

THE TWO DISTILLERIES ON THE SAME HILL

Clarke's charming illustrations from the Jameson pamphlet Elixir of Life
make this arguably one of the loveliest books ever produced on whisk(e)y

new distillery that stands, a mile away, on the waters of the second burn.

Third, peat. The convenient proximity of a peat bog is an economic necessity for a Highland malt distillery.[6] On the north shore of Scapa Flow is Hobister Moor which supplies the peat for the Highland Park distillery. The famous Faemussach Moss, with its inexhaustible peat deposits, contributes something to the distinctive flavour of Glenlivet whisky. For, of course, there is peat and peat. Good whisky is very fastidious in its tastes and demands a peat which is wholly free from mineral impregnations. Even among peats which are not contaminated in this way there is a certain amount of variety owing to differences in the vegetable composition of the fuel.

Fourth, barley. It is a little too late in the day to pretend that all whisky malts are made from barley grown in the district where the distillery is situated. It has even been said that the barley grown and ripened in a warm climate is more dependable for malting than home-grown barley, which may have suffered from rain after it was in the 'stook'.[7] The

---

[6] *Quite simply not a sentence that could or would be written today.*

[7] *East Anglian barley is frequently preferred by maltsters due to its lower moisture content. As a general rule the Scotch whisky industry now sources its barley internationally. Larger brands which once made great play of their use of a particular variety, regarding it as a pillar of their identity, have now quietly abandoned the golden promise of that claim. Some smaller*

proprietor of one well-known Speyside distillery said recently that he had examined whisky made from Californian, Danish, Australian, English and Scotch barley and had not been able to detect any very marked difference between them. Still, he confessed that, if other things were equal, he preferred to use a homegrown barley from the Scots seaboard lying between the rivers Deveron and Ness. The Glenlivet distillery has obtained its barley for generations from the same farms lying in the deep, fertile fields of the Laichs of Banff and Moray, and in this it accords with the practice of the makers of most of the classic whiskies. An inherited skill in raising barley for distillation—for a great deal depends on the degree to which the grain has been allowed to ripen before cutting, and on its garnering and threshing—accounts for the long and close association between farm and distillery which is so common in the North. And, of course, locally-grown barley, even where it is not superior to imported grain, helps to preserve the traditional character of the spirit made from it.

---

producers such as Bruichladdich still source much of their barley as locally as possible and are able to offer expressions made exclusively with Scottish barley. However, it is an undeniable fact that there is simply not enough malting barley grown in Scotland to support the distillers' demands, and whatever the romantic appeal and traditional claims of home-grown barley, in order to sustain current levels of production Scotch whisky will have to continue the use of foreign barley for the foreseeable future.

The whisky districts of Scotland—for Irish whiskey belongs to a separate kingdom and must receive separate notice—are four in number.[8] They are:

1. HIGHLAND MALTS
2. ISLAY MALTS
3. CAMPBELTOWN MALTS
4. LOWLAND MALTS

The first three of these districts share in common the distinction of employing peat to dry the malt in the kiln. But the Islays and Campbeltowns are not to be thought of as mere subdivisions of the Highland area; they are quite independent and distinct units, and the whiskies they make belong to indigenous and inimitable types which have a definite place and function in the orchestra of whisky.

There are, at the time of writing, 122 distilleries in existence in Scotland.[9] Of these ten make Islay

---

[8] *See the earlier note on geographical classifications.*

[9] *Strange as it may seem, I am not going to attempt to give a definitive number for the number of distilleries currently operating in Scotland—largely because the number keeps changing and you will find the current information on the web.*

The answer, if grain distilleries are included, will be 'about a hundred', but there are only seven grain distilleries. Incidentally, the one you can't think of, even if you are a serious whisky enthusiast, is the determinedly anonymous Starlaw near Bathgate where a French group built a large state-of-the-art plant a few years ago.

That means there are more than ninety single malt distilleries, but with the explosion of small craft distilleries it's hard to keep

Campbeltown was home to an important group of distilleries now greatly reduced in number

whiskies; ten are in Campbeltown; eight make Lowland malt; one makes both grain whisky and Lowland malt; nine produce a grain whisky; and by far the greater number, eighty-four, distil Highland malt whisky. In other words, two-thirds of the entire whisky-distilling industry of Scotland is devoted to the manufacture of the Highland malt spirit.

The habit that distilleries have of crowding into a few closely-defined areas whose natural conditions have been proved by centuries of experience to be most suitable for distilling, is well illustrated by

up: the number could well be over a hundred as you read this. And some large distilleries are being built as well, such as Inchdairnie in Fife, but we won't see its single malt until 2028 at the earliest.

However, it is still the case that the Highland region, including Speyside, accounts for the largest number. Sadly, both Campbeltown and the Lowland region have declined in importance since MacDonald was writing.

the Campbeltown malt whiskies. The whole of this important group of distilleries is found in or around the town of Campbeltown on the peninsula of Kintyre, which juts southward from Argyleshire to within a few miles of Northern Ireland. Within the few square miles which make up the whole of this, the smallest of the four whisky areas, a spirit is produced which differs widely from any of the other types of Scotch whiskies. The Campbeltowns are the double basses of the whisky orchestra. They are potent, full-bodied, pungent whiskies, with a flavour that is not to the liking of everyone. Indeed the market for these whiskies is largely confined to Scotland and to the western part thereof. So masterful and assertive are they that the marrying of them to obtain a smooth, evenly-matched blend is an extremely difficult business. Yet, if the full repertoire of whisky is not to be irremediably impoverished the Campbeltowns must remain. As might have been expected in an age when the standardized, anaemic grain-plus-malt are triumphant, Campbeltown distilling has been somewhat under a cloud in recent years. This district has suffered more severely than the others from trade depression. A few years ago it would have been necessary to mention seventeen Campbeltown distilleries, but in the interval the stills of seven of them have grown cold. The ten

Campbeltown whiskies which remain are:[10]

SPRINGSIDE	LOCHHEAD
RIECLACHAN	BENMORE
KINLOCH	SCOTIA
HAZELBURN (KINTYRE)	LOCHRUAN
GLENSIDE	SPRINGBANK

The Islay distilleries, also ten in number, form a province of whisky no less autonomous and distinct than the Campbeltowns. They are sprinkled along the coast of the beautiful island of Islay off Argyleshire. We may call them the violincellos of the orchestra, somewhat less heavy and powerful of flavour than the Campbeltowns, yet perfectly equipped after their insular fashion, round and well-proportioned. More friendly and accommodating than their brothers of Kintyre, it is their glory that they make a good marriage with the Highland malts of the Spey district, bringing out the qualities of their mate without sacrificing any of their own beauties in

---

10 *Even as MacDonald was writing there were further closures in Campbeltown. There were a variety of interrelated factors behind this collapse.*

Today only Springbank and (Glen) Scotia survive from his original list, though Glengyle has subsequently been restored and re-opened. None of these three could be described as a major producer, and Campbeltown today is a pale imitation of its former glory. MacDonald's description of them is poignant: one wonders just how aware he was of the Campbeltown industry's parlous state in 1930. However, writing around forty years earlier, a clearly unimpressed Alfred Barnard described the regional style as 'thin', only 'to be used in moderation [in blends] and never allowed to predominate'.

the process. They give breadth and fullness to the harmony but they do not drown the voices of less capacious instruments. The ten Islay whiskies are:[11]

BUNNAHABAIN	ARDBEG
CAOL ILA	MALT MILL
BRUICHLADDICH	LAPHROAIG
LOCHINDAAL	PORT ELLEN
BOWMORE	LAGAVULIN

Of these ten the most esteemed are probably Caol Ila, Ardbeg, Laphroaig, and Lagavulin, four whiskies with an almost legendary fame. The other day I met a man who during his life as a recruit in the army was kept awake for hours in the night by the prolonged rhapsodies of two Highlanders, men who had nothing else in common in the world but their affection for and praise of Lagavulin.[12]

---

[11] *Lochindaal closed around 1929 and, despite relatively recent plans to rebuild, has never worked since. Only the warehouses survive.*

Port Ellen was closed in 1983 and will not re-open. During the 1980s all of the surviving distilleries were working at a low ebb, if at all, and Islay whisky was out of favour both with the blenders and the consumer. However, the style has enjoyed a considerable revival since then. A small farmhouse-style distillery was opened at Kilchoman in 2005, and there are currently tentative proposals for a further three new operations with the possibility of yet more to be announced.

[12] *Unusually for any distillery, Lagavulin, which celebrated its two hundredth anniversary of legal distilling in 2016, is mentioned by name in a number of Victorian sources. Distilling is said to have begun on the site in 1742.*

The Lagavulin Mash Tun and Mash Man, ca. 1930

On maps in histories of Scotland one often sees a line, called the 'Highland Line', drawn slantingly through the country, north-east from the Clyde estuary in Dumbartonshire through the middle of Perthshire and then veering west and north to include the uplands of Forfarshire, Aberdeenshire, and Morayshire. It runs out to sea about Nairn and reappears at the boundary of Caithness. To the west and north of this line from the sixteenth century onwards was the 'pale' of the Celtic clan system and the Gaelic speech. But the Highland Line on a whisky map of Scotland is kinder to Celtic susceptibilities and includes great stretches of country which have been politically, racially, and linguistically Lowland for long centuries. If a straight line be drawn on the map between Dundee on the east and Greenock in the west it will represent the boundary between Highland and Lowland malt whiskies. Everywhere to the north of this invisible frontier is the dominion of the Highland whisky, its southern outpost in Stirlingshire on the very edge of the line. (And, indeed, as in the case of other frontiers and other outposts, there is some dispute as to which territory is the rightful owner of this distillery, whose product, of less markedly Highland character than the others, appears variously as a Highland and a Lowland malt in different trade lists.)[13]

---

13 *Presumably Glengoyne.*

GEOGRAPHY

The Highland whisky area is made up of one or two fairly distinct sub-divisions, of several scattered units which may be grouped together for convenience under county headings, and of one thronged and important core or nucleus. Ignoring this last for the moment, we shall take the Highland whiskies in rough geographical order, beginning with the most northerly.

Pomona, the main island of Orkney, forms one of the sub-divisions of the Highland area. It has three distilleries, Scapa, Stromness,[14] and Highland Park, at Kirkwall, of which the last is probably the best known. Depending largely, as they do, on imported barley, much of which comes to them from the great grain-growing lands of the Lothians, these distilleries produce a whisky of strong individuality, resembling not the famous Banffshire makes but rather the Aberdeenshire group of Highland malts which are sometimes given the name of 'North Country' whiskies. Highland Park, the best of the three, is one of the small first class, the premiers crus, as it were, of Scotch whiskies. The distillery is the most northerly of all and one of the most ancient, having been founded in 1789 near the site of a bothy kept by one

---

[14] *The Stromness distillery had, in fact, closed by this date but is remembered for its charming publicity materials (see illustration over page).*

Scapa is working today under the ownership of Chivas Brothers. It is notable for operating a Lomond still, somewhat modified.

A delightful promotional postcard for the Stromness Distillery's Old Orkney brand. The distillery closed in 1928 when the town voted to go 'dry'

Magnus Eunson, a famous Orkney smuggler, who was beadle (*Anglicè*—verger) of the local U.P. Church and is said to have concealed his illicit spirits under the pulpit of the Church. When excisemen attended the services, not altogether for pious reasons, Eunson, we are told, used to announce the psalms in tones of exceptional unction.[15] He was a true brother in the spirit of that other illicit but devout distiller who, in reply to the reproaches of his minister, said, 'I alloo nae sweerin' in the still, everything's dune decently and in order.' But it is unlikely that the Orkney clergy found fault with the illegal activities of their flocks. In Hall's *Travels in Scotland* (1807) we read, 'It is a shame that the clergy in the Shetland and Orkney Isles should so often wink at their churches being made depositories of smuggled goods, chiefly foreign spirits.' Indeed, if we owe the still which has given us Highland Park to the convenient blindness of the Orkney ministers, there will be those among

---

[15] *This tale appears in Barnard, but I think it more likely that MacDonald has taken it directly from promotional material issued by the distillery in 1924. The story appears in their little pamphlet,* A Good Foundation, *where we read "when Excisemen attended the services (from motives not unconnected with the Spirit!) he is said to have announced the psalms in tones of unusual unction".*

Such was his haste in working that he has transcribed the date that the distillery was founded incorrectly! A few lines above the sentence quoted from A Good Foundation *it is given there correctly as 1798; MacDonald has 1789. Such are the perils of plagiarism.*

us who will say that churches have been put to worse uses before now.

Caithness, which in the eighteenth century was sending whisky to Skye and the Hebrides, to-day furnishes one whisky, at Wick, where the Pulteney Distillery is found. Sutherland has at Brora a distillery where the famous Clynelish[16] is made. This admirable malt blends exquisitely with the Speyside whiskies. We have Professor Saintsbury's word for it that the finest whisky he ever blended had for its ingredients Smith's Glenlivet and Clynelish. Ross-shire contributes five whiskies (Glen Skiach[17] is not to be found in the latest register): these are Glenmorangie, distilled at Tain, Dalmore and Teaninich from Alness on the northern shores of Cromarty Firth, and Ferintosh (prematurely mourned by Burns) from the other shore. In Glenoran to the north of Beauly is situated the Ord-Glenoran distillery.

Inverness-shire can scarcely be considered a natural territorial division for whisky. It contains several isolated distilleries or groups of distilleries whose products have no particular resemblance to

---

[16] *The original Clynelish distillery mentioned here was renamed Brora in 1969 when the adjacent new Clynelish distillery was commissioned. The original distillery was then operated intermittently until 1983 when it was finally closed.*

[17] *Glen Skiach closed in 1926, as did Ferintosh, bearer of a noble name. Presumably MacDonald was unaware of the closure or assumed it to be temporary.*

one another. At each end of the Great Glen through which the Caledonian Canal passes is a small whisky district. Near Inverness are the three distilleries of Glen Albyn, Millburn, and Glen Mhor; at Fort William, at the southern or western end of the Glen, are the distilleries of Glenlochie and Ben Nevis. But the Inverness whiskies[18] approach the Banffshire type in character, while the Fort William products belong to a West Highland category, in which the fine Skye whisky, Talisker, is also to be remembered. The other Inverness-shire whiskies are more difficult to classify. Tomatin, made on the Findhorn, has a claim to be included in the inner ring of the Highland whiskies, the core to which reference has already been made. Dalwhinnie, styled a Strathspey whisky, is a native of the Central Highlands, born near the bank of that tributary of the Spey which drains sombre Loch Ericht; it is more than seventy miles from the nearest of its fellow Spey whiskies.

Argyle contributes to the West Highland group Tobermory from Mull, the last of the insular whiskies, Oban, and Glenfyne[19], from Ardrishaig. In Perthshire there is a grouping of distilleries along the valleys of the Tay and its tributaries, the Earn and

---

18 *The three Inverness distilleries are all closed and their sites cleared for housing or retail, as is the case with Glenlochie (more usually styled Glenlochy—either way, it is now lost).*
19 *Glenfyne closed in 1937.*

Key to Map of Distilleries

Glen Skiach.†
Ferintosh.†
Ord-Glenoran.
Glen Albyn.†
Millburn.†
Glen Mhor.†
Bunnahabain.
Caol Ila.
Bruichladdich.
Ochindaal.
Bowmore.
Ardbeg.
Malt Mill.†
Laphroaig.
Port Ellen.
Lagavulin.
Glengoyne.
Edradour.
Stronachie.†
Glencoull.†
Glencadam.
Glengarioch.
Ardmore.
Glencawdor.†
Brackla.
Dallas Dhu.†
Ben Romach.
Glenburgie.
Milton-Duff.
Glenlossie-Glenlivet.
Longmorn-Glenlivet.
Linkwood-Glenlivet.
Glenmoray-Glenlivet.
Glenelgin-Glenlivet.
Coleburn-Glenlivet.†
Speyburn-Glenlivet.
Glenspey.
Glenrothes-Glenlivet.
Glengrant-Glenlivet.
Craigellachie-Glenlivet.
Glentauchers-Glenlivet.
Inchgower.
Strathmill.
Aultmore-Glenlivet.

44. Strathisla-Milton Keith.
45. Knockdhu.
46. Banff.†
47. Macallan-Glenlivet.
48. Aberlour-Glenlivet.
49. Benrinnes-Glenlivet.
50. Convalmore-Glenlivet.†
51. Balvenie-Glenlivet.
52. Towiemore-Glenlivet.†
53. Parkmore.†
54. Glendullan-Glenlivet.
55. Mortlach.
56. Glenfiddich.
57. Dufftown-Glenlivet.
58. Tamdhu-Glenlivet.
59. Cardow.
60. Imperial-Glenlivet.†
61. Knockando.
62. Glenfarclas-Glenlivet.
63. Dailuaine-Glenlivet.
64. Cragganmore-Glenlivet.
65. Balmenach-Glenlivet.
66. Glenlivet.
67. Stratheden.†
68. Cameron Bridge.
69. Auchtertool.†
70. Grange.†
71. Glenochil.†
72. Cambus.†
73. Rosebank.†
73a. Bankier.†
74. Littlemill.†
75. Auchentoshan, Duntocher.
76. Gartloch.†
77. Glenkinchie.
78. Kirkliston.†
79. Bladnoch, Wigtown.

The Campbeltown distilleries:
Springside,† Rieclachan,† Kinloch,† Hazelburn,† Glenside,† Loch-head,† Benmore,† Scotia, Lochruan,† Springbank.

†*Distilleries that are now closed*

the Tummel. At Pitlochry the Blair Atholl whisky is made with, close beside it, Edradour of the poetic name; Ballechin is distilled at Ballinluig, Aberfeldy at the charming little town of that name on the upper Tay, Glen Turret at Crieff, and Isla at Perth.[20] All are useful, robust whiskies, somewhat below the highest grade in delicacy.

In Stirlingshire, Glengyle just scrapes into the Highland area (as the best opinion holds), and with Forfarshire we have reached the southern edge of the range of North country malts which extends to Peterhead. The beautiful glens and uplands of the Sidlaws are the cradle of Glencoull, Glencadam, and North Port, the Brechin whisky. Kincardine adds the two Mearns whiskies, Auchenblae and Glenurie, distilled at Stonehaven.

When the Aberdeenshire border is reached on the return northward journey the north country character of the malt is well-established. There are six distilleries within the county area, Lochnagar Royal at Balmoral (one of the two distilleries entitled to use the regal adjective), Strathdee in Aberdeen city, Glengarioch near Old Meldrum, Ardmore to the south of Huntly, Glenugie at Peterhead, and

---

20 *Of the distilleries mentioned here Ballechin (1927), Isla (1926), Glengyle (c. 1925), Glencoull (1929), North Port (1983), Auchenblae (1926), Glenurie, later Royal Glenury (1985), Strathdee (1938) and Glenugie (1983) are all now closed.*

Glendronach at Huntly. The last-named is probably the most distinguished of this group; Professor Saintsbury mentions it kindly. But it seems doubtful whether a Huntly-distilled whisky should not be numbered with the Banffshire family. The county border-line is, in a case like this, of less significance than natural geographical divisions, and Huntly is situated upon a tributary of the Deveron, which rises among the Cairngorms and flows northward through the rich grain-bearing slopes that come down from the Grampians to dip into the North Sea along the east-and-west running coastline that extends from Nairn to Fraserburgh. There would be a case, then, for considering Huntly as within the same geographical area as nourishes the great Spey whiskies. We shall leave it, like Tomatin and Dalwhinnie, with a question-mark behind it.

The real heart of this whole matter of whisky lies in a rough quadrilateral of land, about fifty miles by twenty-five in area and corresponding approximately to the three Scottish counties of Nairn, Moray, and Banff. It would be no true or, at least, no very discerning lover of whisky who could enter this almost sacred zone without awe. It would be a most unimaginative man who could pass along its roads and look on its woods and fields, its pleasant hills and beautiful sea-shores without seeking for analogies in the scenery around him with the various perfections in his favourite whiskies. The tract has been very decisively marked off by nature, as if on purpose to contain and guard some hallowed mystery. To the north there is the sea, a magnificent sea, with already something in it of the Arctic chill and the Arctic eeriness. To the west is the river Nairn, to the east the Deveron.[21] On the north, enclosing the region with a superb sculptured wall, are the two mountain masses of the Cairngorms and the Monadhliath mountains, between which the powerful Spey thrusts itself like a spear into a closing door. Five rivers water the area, passing northward (more accurately N.N.E.) to the sea: Nairn, Findhorn, Lossie, Spey and Deveron. It is one of the most fortunate areas in all Britain, in its climate, in its scene, in the fertility of its soil and the grandeur of its pine woods, in the physical dignity and mental poise of its people. Here Scandinavian, Celt, and those elder peoples who were before the Celt and seem likely to live as long as he, have met and mingled with the happiest of results. Such is the inner sanctuary, the fountain-head, the Ark of the Covenant of whisky.

Looked at from the purely quantitative aspect, this compact area has an importance which can easily

---

[21] *A new distillery, MacDuff, was built by the River Deveron in 1962–63 to a then radical design. It continues to operate to this day.*

be measured. Out of the 122 Scottish distilleries, eighty-four are in the Highlands; of these eighty-four Highland distilleries forty-five are situated in this single district. Yet it must be noted at once that the names of the whisky districts as they are shown in the registers of the industry do not take cognizance of the unity of this district. Indeed it is impossible to discover any reasonable geographical basis for the district-names assumed by different distilleries, apparently by caprice. No amount of research can determine why some of these are called 'Strathspey' whiskies and others 'Speyside'. Both are equally beside the Spey and in Strathspey. And it is not as if the names were interchangeable; they are jealously guarded. The different trade-designations given to the whiskies belonging to this area, the central malt whiskies, are as follows: Speyside, Strathspey, Nairn, Elgin, Forres, Dufftown, Rothes, Knockando, Banffshire, Keith, Glenlivet.

The simplest and more logical sub-division of the area would be into five districts, roughly corresponding to the courses of its five principal rivers. Thus the Nairn belt would include the two whiskies at present known as Nairn whiskies; the Findhorn belt would take in the three Forres whiskies and one of the whiskies at present grouped under Elgin; the Lossie area would comprise the six remaining Elgin whiskies; the Spey whiskies would number the present Strathspey and Speyside, and in addition the Rothes, Knockando, Dufftown, and Glenlivet brands, twenty-seven in all; while the Deveron area would include those whiskies which now are called Keith and Banffshire, six. But in this matter, as in others more important, the conservatism of the distillers will probably prove unassailable.

Of the two Nairn whiskies, Royal Brackla and Glencawdor, the former is the more important. It is, in fact, one of the dozen or so best whiskies made in Scotland. In this case the barley used comes from the counties of Moray and Nairn and the water from springs high in the Cawdor Hills. It acquired the epithet 'Royal' early in its career. In 1828 an advertisement in the *Aberdeen Chronicle* announced that Captain Fraser[22] 'has made an arrangement to have this *much admired spirit* sent up by land, when a regular supply can be had weekly'. But by 1835, as an advertisement in *The Morning Chronicle* of London informs us, 'Brackla Highland Whisky' had become 'Brackla, or, The King's Own Whisky'.

---

22 *Captain Fraser was a notoriously tricky character to deal with, being fined more than once for breaching the distillery laws. Despite this, the distillery held two Royal Warrants by 1838. He is mentioned in distinctly unflattering terms in Joseph Pacy's* Reminiscences of a Gauger *(1873), though Pacy himself appears to have been something of a martinet.*

Today the distillery, much remodelled, is operated by John Dewar & Sons and most of the output goes to that company's blends.

'His Majesty, having been pleased to distinguish this "by his Royal Command to supply his Establishment", has placed this Whisky first on the list of British Spirits, and, when known, should in truth be termed "The Drink Divine"—only to be had of the Importers, Graham and Co., New-road, facing the Mary-labonne workhouse.'

The whiskies at present classified as Forres brands belong to the Findhorn group I have suggested. They are three in all: Dallas Dhu, Ben Romach, Glenburgie.

Elgin, on the Lossie, gives its name to seven whiskies of which one, Milton-Duff, seems to belong by geography rather to a Findhorn group. The other six are Glenlossie-Glenlivet, Longmorn-Glenlivet, Linkwood-Glenlivet, Glenmoray-Glenlivet, Glenelgin-Glenlivet, and Coleburn-Glenlivet.[23]

Of the whiskies belonging to the famous Spey basin those classified as members of the Rothes district account for three, Glenspey, Glenrothes-Glenlivet, and Glengrant-Glenlivet. Dufftown includes five notable whiskies, Convalmore, Balvenie, Glendullan-Glenlivet, Mortlach, and Glenfiddich. There are twelve distilleries which own no more exact geographical title than 'Speyside'. They are

---

23 *With the exception of Dallas Dhu, now a museum, and Coleburn, closed in 1985, all of these distilleries remain open.*

Craigellachie-Glenlivet, Glentauchers-Glenlivet, Imperial-Glenlivet, Macallan-Glenlivet, Parkmore, Dufftown-Glenlivet, Daluaine-Glenlivet, Cragganmore-Glenlivet, Benrinnes-Glenlivet, Speyburn-Glenlivet, Towiemore-Glenlivet, and Glenfarclas-Glenlivet. 'Strathspey', a designation differing from 'Speyside' in some subtle way which no inquiry and no consultation with mere maps can elucidate, is given by three distilleries as their *pays d'origine*. In addition to Dalwhinnie, situated to the south of the Cairngorms in Badenoch and already noticed, there are Balmenach-Glenlivet and Aberlour-Glenlivet. The three Knockando whiskies are Tamdhu-Glenlivet, Cardow, and Knockando. The last-named, like the Rothes distillery, Glenspey, is owned by Messrs. W. & A. Gilbey Ltd., who use its product for the blending of their proprietary whiskies. The Banffshire malts, which may also be included in the Spey group, are Inchgower, a coastwise whisky distilled at Buckie, Banff, distilled in the county town, and Strathmill, Strathisla-Milton Keith, Knockdhu, all of which are made in or near Keith, which however is recognised as a separate district by only one whisky, Aultmore-Glenlivet.[24]

---

24 *Of the distilleries mentioned here, Convalmore (1985—an especially sad loss), Imperial (closed 1998 and demolished but a new distillery, Dalmunach, was opened on the site in 2013), Parkmore (1931, though the warehouses remained in use until at least the late 1990s and possibly longer)*

Only one name has still to be mentioned, but it is a name of supreme importance. There is a small glen in Banffshire watered by a stream named the Livet, a tributary of the Avon which is itself a tributary of the splendid Spey. The pure waters of this mountain burn which tumble 1,200 feet down the steep sides of Cairngorms do something more than drain the beautiful glen, they also form an essential ingredient in the whisky for which Glenlivet is renowned. Glenlivet has for long decades been almost a synonym for the finest of Highland whisky:

> Glenlivet it has castles three
> Drumin, Blairfindy, and Deshie,
> And also one distillery
> More famous than the castles three.

and Towiemore (1930/31) are all permanently closed.

The famous Glenlivet distillery, ca. 1925

Product of a district of which a century ago it was said, 'everybody makes whisky and everyone drinks it', Glenlivet, 'the real Glenlivet', won an early place in the literature of whisky. Sir Walter Scott in St. Ronan's Well, that locus classicus for so much relating to the Scottish kitchen and the Scottish cellar, makes Sir Bingo treat the Captain and the Doctor to a cordial prepared in the wilds of Glenlivet. The Captain's verdict was as follows: 'By Cot, it is the only liquor fit for a gentleman to drink in the morning, if he can have the good fortune to come by it, you see.' 'Or after dinner either, Captain,' added the Doctor. 'It is worth all the wines of France for flavour, and more cordial to the system besides.' Still more eloquent and enthusiastic is the eulogy which Christopher North[25] puts into the mouth of James

---

25 *Christopher North was the pen name of Professor John Wilson, whose* Noctes Ambrosianae *(Ambrosian Nights) appeared between 1822–1835 in* Blackwood's Magazine. *These consisted of a series of fictional observations purported to have been made by a cast of characters in Ambrose's Tavern, Edinburgh (also imaginary but said to be based on Tibbie Shiels Inn by St Mary's Loch).*

The Ettrick shepherd is indeed James Hogg, one of an Edinburgh literary circle and today best remembered for his novel The Private Memoirs and Confessions of a Justified Sinner. *He was less than amused by the satirical portrayal of his alter ego though it seems that he did on occasion enjoy Glenlivet with some enthusiasm.*

As a minor curiosity there was once a real Ambrose's Tavern in Edinburgh, on Gabriel's Road near the Register Office, though it was not the scene of North's stories.

Hogg, the Ettrick shepherd, and one of the most discerning of the old Scottish school of gourmets: 'Gie me the real Glenlivet, and I weel believe I could mak' drinking toddy oot o' sea water. The human mind never tires o' Glenlivet, ony mair than o' caller air. If a body could just find oot the exac' proper proportion and quantity that ought to be drunk every day, and keep to that, I verily trow that he might leeve for ever, witout dying at a', and that doctors and kirkyards would go oot o' fashion.'

In a later age, a less august authority, the frivolous but ingenious author of *The Massacre of Macpherson*, a ditty without which no Scottish students' revel is (or was) complete, celebrated the malt in a stanza:

> Phairson had a son
> Who married Noah's daughter,
> And nearly spoilt the flood
> By trinking up ta water.
> Which he would have done—
> I, at least believe it—
> Had ta mixture been
> Only half Glenlivet.[26]

---

[26] *This humorous poem, mocking the stereotype of the Highlander and his pronunciation of English, is by William Edmondstone Aytoun (1813–1865), an Edinburgh lawyer and academic. According to the National Library of Scotland it appeared in print as a broadsheet ballad published by*

But the most striking tribute to Smith's Glenlivet[27] is not a literary one. There are at the present moment no fewer than twenty-six distilleries which use the name Glenlivet in a hyphenated form. These distilleries are found over an area of about three hundred square miles. There was some point in the old joke 'Glenlivet is the longest glen in Scotland', especially since the hyphenation was not strictly observed at one time. So grave was the inconvenience to the makers of Glenlivet that in 1880 Colonel John Gordon Smith, the proprietor of the distillery, took the question to law, and it was decided that only the Glenlivet Distillery was entitled to style its product 'Glenlivet' without qualification.

The roll of the Highland malt whiskies of Scotland is now complete. It is an invidious as well as difficult task to single out any for special

---

the Poet's Box, Glasgow. NLS date this to 1880–1900, which seems late as by 1891 it had appeared in the Scottish Student's Book of Song, *which remained in print until at least 1939. MacDonald presumably encountered it during his student days, but it is still recorded as being sung by Scottish students at Cambridge University in the early 1950s. The verse concludes:*

> *'This is all my tale*
> *Sirs, I hope 'tis new 't ye!*
> *Here's your fery good healths*
> *And tamn [alt. hang] ta whusky duty.'*

[27] *Glenlivet perhaps needs no further praise from literary men but its qualities were also noted in the technical literature. For example, in* The London Dispensatory *of Anthony Todd Thomson (1837) he baldly states, 'The best Scotch whisky is Glenlivet'.*

recommendation, yet a list of twelve names can be made up which will probably win all but universal acceptance as representing Highland whisky at its most distinguished. It is as follows:[28] Glen Grant, Highland Park, Glenburgie, Cardow, Balmenach, Royal Brackla, Glenlossie, Smith's Glenlivet, Longmorn, Macallan, Linkwood, and—. But the twelfth place I decline to fill, being unable to decide, even after prolonged spiritual wrestling and debate, whether Talisker or Clyne Lish should be honoured. It is a problem which some of my readers may, in any case, think it desirable to settle for themselves.

The Lowland malt whiskies of Scotland are, with one exception, distilled in a narrow belt of country bounded on the north by the 'Highland line' already mentioned and on the south by a roughly parallel straight line running from Largs on the west to St. Abb's Head on the east. Within this area are seven distilleries, Stratheden in Fifeshire, Bankier in

[28] *Fashions change: today such a list would undoubtedly include one or more Islay whiskies and probably Springbank from Campbeltown. Both Glenburgie and Royal Brackla, which have been extensively remodelled and are now used virtually exclusively for blending, would be unlikely to appear, and the same is true of Glenlossie and Balmenach; both are now rather obscure. Cardhu might find a place, but I rather doubt it. Amongst the notable absentees we might list Glenfiddich, The Balvenie, Aberlour and Glenfarclas. MacDonald is here rather careless of his own geographical boundaries; neither Highland Park nor Talisker could properly be considered Highland whiskies, fine though they undoubtedly are.*

Stirlingshire, Rosebank at Falkirk, St. Magdalene at Linlithgow, Glenkinchie at Haddington, Littlemill and Auchentoshan, Duntocher, both in Dumbartonshire. There is also the Yoker distillery in Glasgow making both malt and grain spirit. Far removed from this central Scottish plain, well-peppered with distilleries, is the Bladnoch distillery at Wigtown, making a whisky which, like its situation, stands somewhat apart from the rest of the Lowland malts.[29]

The ten distilleries making a grain spirit are all to be found in the industrial Lowlands of Scotland. They are Carsebridge, Cameron Bridge, Glenochil, Port Dundas, Adelphi, Strathclyde, Yoker, Gartloch, Caledonian and North British. Glasgow and Edinburgh account for six of them. The best of these grain distillations, as established by the experience of the blenders, are Caledonian, Carsebridge, Cameron Bridge, and North British. It is maintained that these four whiskies have each a definite individuality—but this is a heresy that rouses the average malt distiller to passionate rage.[30]

---

[29] *With the exceptions of Glenkinchie, Auchentoshan and Bladnoch, all of these distilleries are now lost. Rosebank and St Magdalene in particular are sadly missed.*

[30] *MacDonald had presumably not tasted grain whisky. There is indeed a marked difference between the various makes; possibly not as wide as between single malts but clear and discernible nonetheless.*

 Much has changed since this list was compiled. Cameronbridge, Strathclyde and North British remain open and new grain distilleries have been

Ireland is properly considered as a separate kingdom in the world of whisky—or whiskey as we must now begin to call it. For not only has it a geographical claim to individual treatment, but there are many characteristics which mark it off from its Scottish brothers. For instance, the process of manufacture is more elaborate, there being three distinct distillations before the finished spirit is collected. For the most part it is made from a mash of barley and barley malt and not from barley malt alone, though there are some exceptions to this rule, of which old Buskmills whiskey is the most notable. Irish whiskey has somewhat fewer and less delicate varieties of flavour than Scotch (though I have known Scotsmen who preferred it to their native spirit), its range is definitely narrower, and, in the opinion of Professor Saintsbury and other authorities, it is better as a single or 'self' whiskey than in blends. But a fine pot-still Irish whiskey is a magnificent thing indeed, even to those who are not prepared to go as far as one expert and affirm that it is possible to mistake Irish whiskey for cognac. It should be noted that the Irish variety needs more keeping than Scotch, being best after ten years or so in the cask.

---

constructed at Invergordon, Girvan, Loch Lomond and Starlaw (Bathgate) since MacDonald was writing, making seven active plants at present.

Grain whisky is now actively promoted in its own right, most notably by a former English footballer and 'fashion icon'. What MacDonald would make of this is a line of enquiry perhaps best not pursued.

There are twenty-one Irish distilleries in operation at the present time. This number, negligible when compared with the Scottish centres of manufacture, suggests strongly that Irish whiskey has failed to hold its place in the world market,[31] and possibly that, at home, it has fought a losing battle against 'poteen', which today threatens to become a source of supply for American bootleggers. In 1800 there were 124 distilleries in Ireland. It is unfortunately true that the name of Persse of Galway and other distilling firms of equal repute are no longer to be found in the semiofficial trade lists, but it is still possible to find Irish whiskey in a wide enough range of makes to satisfy most tastes.

The (legal) distilling of Ireland is confined rather narrowly to a few, mostly urban, districts. Eleven distilleries (more than half of the total) are within the boundaries of Co. Down, Co. Antrim and Co. Derry in the north-east corner of the island. Of the remainder, five are in Dublin, all of them making whiskey in pot-stills. These are John Jameson's Bow

---

[31] *Irish whiskey had indeed failed to hold its place in the world market and, regrettably, had a lot further to fall in the years following publication. Once arguably the greatest producing nation in the world, Ireland is only now recovering its position with a spirited revival of production being reflected in a number of distillery openings, both large and small.*

The five Dublin distilleries that he mentions have now all been closed. Until very recently there was no distillery in Dublin, a tragic state of affairs, but recently two have opened in the historic Liberties district.

Street Distillery (founded in 1780, curiously enough by a Scotsman, member of an Alloa family, and the sheriff clerk of Clackmannan), the D.W.D. or Jones Road Distillery of the Dublin Distillery company, John Power & Son's John's Lane Distillery, the Marrowbone Lane distillery (William Jameson's) and the Thomas Street distillery (George Roe's) both of which are now owned by the Dublin Distillery Co. In the very heart of Ireland are John Locke & Co.'s Kilbeggan Distillery (Co. Westmeath) and B. Daly & Co.'s Tullamore Distillery (King's Co.). At Cork are the two distilleries of the Cork Distilleries Co. Ltd. and at Bandon, Co. Cork, Allman & Co.'s Bandon Distillery.

Of the twenty-one Irish distilleries, three make a pure malt whiskey, the famous 'Old Bushmills' Distillery at Bushmills, Co. Antrim, near Portrush, and the Coleraine and Killowen Distilleries, both in Co. Derry. A grain spirit is distilled at Abbey Street, Londonderry, and at Avoniel and Connswater (with pot-still), Belfast. It will be seen that all these are situated in the three north-eastern counties. All the rest make a pot-still whiskey. There are two distilleries at Comber, Co. Down, both owned by the same firm, two at Londonderry, Limavady and Waterside, and one in Belfast, the Royal Irish Distillery, where Dunville's well-known whiskey is made.

Four English distilleries—three in London: Wandsworth, Hammersmith, and Three Mills;[32] and one in Liverpool, Bankhall—produce grain spirit.

There is, fortunately for the reader's patience, no longer any need to discuss the geography of whisky in America (though Canadian whisky distilled from rye, maize, etc. should be mentioned as a competitor with the less distinctive Scotch blends). It may be presumed that the ingenuity of the American people is finding its own remedies for the perils of Prohibition[33] and bootleg whisky, though, as a matter of fact, having observed the effects on the health of the population of several hard-bitten Scottish cities, of a few 'dumped' shiploads of Kentucky rye whisky, I am inclined to doubt whether the present régime can hold much greater terrors than the old. Prohibition has added two more names to the nomenclature of whisky: 'Squirrel' whisky, so called because it induces in its devotees an irresistible desire to climb trees, and 'Rabbit' whisky which creates an impulse to leap and run.

---

[32] *All four are gone and it would be a generous man who would suggest that they are greatly missed. Of course, gin is distilled in England in considerable quantities and very successfully so. However, the UK's best-selling gin, Gordon's, is distilled wholly in Scotland.*

English whisky is now made in Norfolk, London and the Lake District and finding some success, albeit the quantities are very small.

[33] *As regards* Whisky *and Prohibition see my notes on the publishing history in the Appreciation.*

But the whisky student has already enough names to remember without troubling his memory with the zoological brands of the United States. As an aid in his attempts to remember the classic whiskies, the crude doggerel which is here appended may have some value, although, being the work, as I understand, of a Sassenach poetaster, it has taken some liberties with Celtic pronunciation which may excite the anger or derision of the Gael.[34]

RHYMED GUIDE

to the Highland, Islay and Campbeltown malt whiskies of Scotland

Name we first the brands that rule in
Islay in the Western seas:
Bruichladdich, Lagavulin,
Bunnahabain, and Laphroaig.
Once I (lucky fellow!) fell in
With a man who had Port Ellen!
Though, indeed, as good as these
Is Bowmore or Caol Ila,
Celtic witch and arch-beguiler,
Ard Beg, Malt Mill. And I shall
Surely drink more Lochindaal.

---

[34] *I have been unable to trace this poem elsewhere. Perhaps it was MacDonald's own and, conscious of its limited literary merit, he preferred to cloak his authorship in anonymity.*

Last port seen by westering sail,
'Twixt the tempest and the Gael,
Campbeltown in long Kintyre
Mothers there a son of fire,
Deepest-voiced of all the choir.
Solemnly we name this Hector
Of the West, this giant's nectar:
Benmore, Scotia, and Rieclachan,
Kinloch, Springside, Hazelburn,
Glenside, Springbank, and Lochruan,
Lochhead. Finally, to spurn
Weaklings drunk and cowards sober,
Summon we great Dalintober.

Children of the Highland hills,
Products of the Highland stills,
Now's no hour to ponder faults,
Toy with test-tubes, sniff at malts,
Open-chested must we sing:
Away with care—the drink's the thing!
Fearing neither sir nor madam,
Praise we Dufftown and Glencadam.
Wanderer over hill and moor,
Weary, welcomes Edradour,
Purchasing new strength to loin
With Glendronach or Glengoyne,
Glenlochie, or ripe Strath Dee,
Cragganmore and Benachie.

GEOGRAPHY

Pious priest at mass or matin,
'Mid the murmur of his Latin.
Thinks of Mortlach or Tomatin,
Sinning so, but is there any
Sin in dreaming of Balvenie,
Brackla, Millburn, or Glenfiddich,
Cardow, Banff, or Teaninich?
Sailor after months of sailing,
Fishing, yachting, cruising, whaling,
Hears the joyous cry of 'land oh!'
Thirsts at once for choice Knockando.

Let the magistracy glower,
Let the law put forth its power,
He will drink the good Inchgower,
Tamdhu, Parkmore, Aberlour,
And damnation to the finny
Tribes of ocean in Dalwhinnie,—
Drink until the stars go out.
Not for us such deep-sea bout.
Quiet tipplers in our class
Are content with Glenfarclas,
Nor does fancy with us soar
Far beyond sound Convalmore,
Oban, Coleburn, or Dalmore,
With mayhap a straying wish
Towards Glen Elgin or Clyne Lish.
Hopeful nephew bound to see

WHISKY

Wealthy and repulsive aunt
(Shadows of a legacy)
Should equip him with Glen Grant,
He will find the interview
Smoother sailing on Knockdhu.
When debate grows over-heated,
Chairs thrown down and men unseated,
To restore both law and order
Bring in Dallas Dhu, Glen Cawdor,
Speyburn, Longmorn, or Strathmill.
Quick the tempest will be still
And sweet reason reign again
With the flow of Dailuaine.
If an angel unawares
Your domestic table shares,
You will not be wrong to give it
Tumblers of the real Glenlivet!
Serious poets, short of rhymes,
As we all may be betimes,—
For *ars longa, vita brevis*—
Woo the muse with good Ben Nevis,
Though the wench will come no less
For Glengarioch or Stromness,
Scapa or fine Highland Park,
Lighteners of Orcadian dark.
Men will talk most brilliant bosh
On a diet of Ferintosh,
Argue, with emphatic oaths,

Black is yellow on Glenrothes,
Prove that four and four make nine
If encouraged by Glenfyne,
And, in paradoxic fury,
Square the circle with Glenurie.
Converts have been made, they say,
To some quite grotesque belief
By Strath Isla and Glenspey
And Glenturret (made in Crieff).
Cunning preachers rope the sullen
Heathen folk in with Glendullan.
In melée or collieshangie
Glentauchers or Glenmorangie
Timid mortals will inspire
With a high heroic ire,
Though their sudden fits of wrath'll
Quickly pass before Blair Atholl.
Leaders of the hopeless charge,
Rallying for one assault more,
Should have come equipped with large
Flasks of Pulteney or of Aultmore,
Or at least another score
Liquors veterans will think good:
Isla, Ben Romach, Glen Mohr,
Balmenach, Glenburgie, Linkwood,
North Port, Angus-reared at Brechin,
Aberfeldy or Ballechin.
While the vanquished in the fray,

Fleeing to the nearest bar,
Counsel take with Auchenblae,
Comfort seek in Lochnagar,
And, when human courage fails,
Stronachie the foe assails.
Scholar, drinking with a lout,
Knocked his boon companion out,
Bawling egotistically, 'Shall an
Embecile enjoy Macallan?
Craigellachie and Imperial
Are designed for souls aetherial!'
Sad that academic rage
Should pollute my peaceful page;
Class and faction I abhor on
Towiemore or Ord-Glenoran;
Ragged cap and top-hat glossy
Meet as equals on Glenlossie,
Bury hatchets in a hurry
In Glenugie or Glenmoray,
Talisker or Milton-Duff
(Damned be he cries, 'Hold, enough!')
Rounding off at last the story
(Highland section) put we Finis
With Glen Albyn, Tobermory,
Glenglassauch, and Benrinnes.

V

JUDGING, PURCHASE, AND CARE

THE EARNEST student of whisky who has learned that there exist only a hundred and twenty-two Scotch whisky distilleries and twenty-one Irish may be inclined to think that he has exhausted the subject and that it now only remains to transfer his geographical and technical knowledge to the field of practical experiment. If so, a sad disillusionment awaits him. He has left out of account the ingenuity of the blenders who, he will speedily discover, are far more important persons in the commerce of whisky than are the distillers. The measure of their power may be set down with a mathematical exactitude by reference to the current *Directory of Whisky Brands and Blends*,[1] which reveals the astonishing fact that there are no fewer than 4,044 brands of whisky (3,428 Scotch, 487 Irish, and 128—crowning horror!—Scotch-and-Irish) which have acquired by usage an established name and proprietorship. This is what

---

[1] *Presumably a trade directory such as* Harper's. *Scotch and Irish blends have, thankfully, largely disappeared, though from time to time a new one is attempted. At this time the probability is that these were an opportunistic attempt to use the large stocks of Irish whiskey which had built up but proved near impossible to sell in their own right as the Irish industry declined.*

the resourceful blenders have done with the products of those hundred and forty-three distilleries! It has to be admitted, too, that there are remarkably few signposts to guide the would-be buyer of whisky through the labyrinth of this senseless multiplication. There are hundreds of whisky-bottle labels bearing 'Finest old liqueur', 'a blend of rare old Highland malts', 'matured in sherry casks', 'the finest Scotland makes', and suchlike heartening but uninformative legends, but there are remarkably few which state what is the age of the youngest constituent in the blend, or what is the proportion of grain spirit to malt whisky, or in what distilleries the ingredient malts were made.[2]

There seems, in fact, to be something rather like a conspiracy of silence among the proprietors of the different brands of whisky, a conspiracy to prevent the consumer from knowing what he is drinking. There are even cases where the wording of the label appears to be definitely misleading. Thus for example, it may be taken as axiomatic that no Scotch whisky bought by the bottle is a single whisky. Yet there are

---

[2] *Much of this could be written today—and indeed the current campaigns for 'transparency' are anticipated by MacDonald. His extraordinarily prescient commentary illustrates the book's continuing relevance, compelling the attention of today's whisky lovers, though one cannot escape the irony, presumably unintentional, of a pseudonymous author calling for transparency.*

instances where the name of one famous distillery is prominently displayed on the label. The obvious inference is that the contents of the bottle consist entirely of the malt spirit of that one distillery—and this inference is, of course, normally without foundation. Again there is a deplorable looseness in the use of terms like 'fine liqueur whisky'. What does the expression mean? Popularly it is supposed to indicate a whisky containing a slightly higher proportion of alcohol than usual. If this is so—and there is a wide difference of opinion in the trade on the subject—then it is no doubt a sufficient reason for the additional charge of half-a-crown or what not that the consumer is asked to pay. But it would be infinitely more satisfactory to know that the half-crown was buying not so many units of alcoholic content but an extra year or so of maturation, or a higher percentage of one or the other of the first-rank malt whiskies in the blend. After all, whisky is not firewater, and whisky-drinkers are not yet mere soakers, in spite of the scant attention paid to their enlightenment by the trade. It is not without extreme circumspection that one dares to address a few words of mild criticism and suggestion to a body of men so assured of their own commercial acumen as those who compose the whisky trade, but the present status of whisky may perhaps incline them to hearken to suggestions from a disinterested outsider.

The most urgent need of the industry would appear to be some form of trade legislation (carried out, of course, by a trade association) designed to secure a closer definition of trade terms. For example, the words 'finest old liqueur' should not be at the mercy of any unscrupulous dealer with a job lot of grain spirit to dispose of. There ought to be a specific minimum strength and also a minimum age in the cask attached to the phrase. It might be possible, too, to establish a distinction between 'fine old liqueur' and 'finest old liqueur'. Again, it should not be permissible to describe a brand as 'a brand of rare old Highland whiskies' unless there is a recognised minimum proportion of Highland malt present. Leaving considerations of commercial morality aside, some such definitions seem to be recommended by expediency. The present chaotic condition of the trade is affecting the good name of whisky and is tending to array against it ever stronger forces of snobbery and contempt. The present policy of mystification only arouses the suspicion and wrath of the intelligent consumer and gives the unscrupulous proprietor an unfair advantage over his honest rivals. In the interest of the trade, as a whole, the far-sighted plan would seem to indicate a control of the trade's adjectives.

There are other matters which might be recommended, if not as suitable for internal legislation, at least as worthy the attention of individual proprietors. Thus each label on a whisky bottle ought to bear the names of the malt whiskies (grouped as Highland, Islay, Campbeltown, and Lowland) in the blend, and the exact percentage of grain spirit contained in it. In addition, it should state the number of years and months that the blend and each of its constituents has matured in cask.[3] This will seem a somewhat drastic proposal, but the sound whiskies would only gain by it. After all, no one in his senses would dream of buying a bottle of wine if he did not have some guarantee of its age. There are, also, certain penalties attached to falsifying marks of this kind. And why should the buyer of whisky be less intent than the wine-drinker on getting a fair deal? It is not as if age were less important in whisky than in wine. The substitution of a variety of patent metal 'caps', stoppers, and what not for

---

[3] *Nearly ninety years on from this book's publication, the question of age statements on bottles of whisky remains a controversial one. As the industry confronts the consequences of its own unanticipated global success stocks of aged whisky have run short and marketers have responded with the introduction of so-called NAS ('no age statement') whiskies.*

Space does not permit a discussion of this vexing topic, which has been the cause of much controversy on whisky blogs, in social media and elsewhere. It is probable that the trend to NAS whiskies will accelerate, not least for the freedom this gives the producer in formulating the final product.

the familiar corks is a 'reform' which in my opinion has been taken without due consideration. The cork helped to soak up the last traces of alcohols which ought to have been lost in the cask. And it was an insurance against contamination from outside. But are the new caps?[4]

In the meantime, while we await a change of heart—or of head—in the whisky proprietors, what are we buyers of whisky to do? We may give up the fight and fall back on the counsel of despair which dictates the careless command 'Bottle of Scotch, please' and trust that plenty of soda will drown the worst effects of our surrender. Or we may ask for one of the more loudly advertised concoctions, acting on the assumption that what is said in big type and said often enough cannot be false, or on the observation that, if we do not get a very good whisky thereby, we shall not get a very bad one. And there are some brands of whisky which are very bad indeed. But this policy is again something in the nature of a surrender, and, while removing us out of the range of definitely lethal whiskies, it will not take us very far along

---

[4] *A number of producers will assert, privately, that the screw cap closure ('ROPP' in trade jargon) is greatly superior to the cork stopper, which today is often preferred for marketing reasons. However, MacDonald is presumably referring to driven cork closures and few would wish to return to that awkward and inconvenient system with its ensuing risk of TCA contamination.*

the path leading to the higher joys of usquebaugh.

A more adventurous programme, which may involve some disappointments on the way but is certain to contain more than one delightful surprise and to have a happy ending, would begin with the happy traveller leaving the beaten track of the more-advertised brands (remember, they are intended for the great public and therefore have usually been emasculated by over-doses of grain spirit). He is now in an unmapped terrain with only his mother-wit, his nose, and his palate to guide him. The first will take him either to a wine and spirit merchant of unimpeachable repute or to one of those lesser-known brands such as Macdonald & Muir's Highland Queen, which supply some definite information as to origin, age, etc., on their labels. If he should go to a good merchant his exploration will in all probability have begun well. Let him ask for the firm's own blend; it will surprise him what pleasure will be shown on the other side of the counter at the advent of one who does not come bustling in with a loud demand for any one of those brands that are household words. Perhaps he will be unlucky—not all wine merchants take this matter of whisky with the gravity that it deserves. But on the other hand, he may (supposing him to be in London) come upon Justerini and Brooke's 'Club' Whisky or Hedges

& Butler's 'Coronation Vat' or Chalie, Richards, Holdsworth's 'Old Matthew' or any of some scores of excellent blends.⁵ If he take the more speculative course of independent investigation among the

---

5 *Justerini & Brooks is today part of Diageo. Their Club brand was created by Andrew Usher; today they are known for J&B Rare, the blend for which was first developed by another renowned blender Charles Julian around 1930.*

Hedges & Butler was established in 1667 and were once a significant wine and spirit merchant. The whisky brands are owned today by Ian Macleod Distillers.

Chalie, Richards, Holdsworth & Co. were London wine merchants, first

Abbot's Choice and Crawford's Three Star are both today Diageo brands. Crawford's Three Star is available in limited quantities in South Africa and some Latin American territories. However, it appears that Abbot's Choice has not been bottled for some years and its future must be in doubt.

lesser known proprietary brands, he may have the luck to hit upon one of the better whiskies such as 'Abbot's Choice' or Crawford's 'Three Star'. But even if he is not so fortunate in the beginning, he should persist. All the time he will be training his palate and discovering, if not new paradises, at any rate the adumbrations of them. For it cannot be said too often that whisky is not the uninspired humdrum thing that a visit to the average public-house would lead one to suppose.

This matter of the public-house supply of whisky is one that ought to be taken up by an organized and determined public opinion. For in most public-houses you are offered the choice of three (sometimes, with luck, four)[6] well-known brands—and a request for something different is apt to evoke an exhibition of that frosty and supercilious manner in which the British bar-maid excels. The adventurer who is not discouraged by these Arctic divinities nor disheartened by one or two almost inevitable mishaps will be richly rewarded. He should get into the habit of tasting whisky, a practice which, odd as it sounds, has now quite gone out of

---

established in 1775. By 1930 they had premises on New Bond Street but the business seems to have closed shortly after the end of the Second World War.
6 *Here, at least, the situation has improved; the drinker faced with a public house offering just four whiskies would be well advised to take their custom elsewhere!*

fashion. He will find that soda-water has the effect of destroying any distinction a whisky may possess and of reducing the complex and subtle massing of flavours in a fine blend of Highland malts to the level of an ordinary grain-plus-a-little-malt combination.[7] Let him eschew the enticing 'splash', at least until he has put each new whisky to a severe test. Taking a small wine-glass, preferably one with a generous belly and a narrow rim, let him warm it slightly with his hand and then pour in a little of the whisky. Having assisted the process of evaporation with a gentle rocking motion of the glass, let him sniff the vapour at the rim. It should be mild and yet potent, round and 'warm', with no trace of the objectionable acridity of raw spirit. Then he should take a sip—only a drop or two—and allow it to remain in the mouth for a few moments. There ought to be no harshness in the liquor's assault on the palate; 'kick' only indicates a young or badly-mannered spirit. It should be gentle, with nothing of that 'mineral' taste about it which causes all but hardened and careless drinkers to shudder a little and contract the facial muscles; and it ought to possess a smooth, elusive, and varied flavour in which it is difficult to distinguish a dominant constituent. Having performed this

---

[7] *Sound, practical advice which has stood the test of time, especially the recommendation to avoid soda water.*

operation, the experimenter should add some water to the remainder of the spirit and drink thoughtfully; it will be found better to use soft water for diluting, where this is possible, but it is a refinement which the beginner can spare himself.

Having satisfied himself by this simple test that he has encountered a superior whisky, the student should acquire a bottle in order, by a more prolonged association, to determine whether he and the spirit suit one another. He may, for example, find—and this will apply especially to men leading a sedentary life—that a generous content of Highland malt whiskies has the effect of making him liverish.[8] He will then be well advised to keep the blend for special occasions, buying for ordinary use a whisky with a somewhat larger grain admixture.

The most important thing about whisky—its place of origin and malt-content apart—is its age. The student should never buy a brand whose age is not stated and guaranteed. He ought to do more, for there is no reason why he should not take a hand in improving the quality of the whisky he buys. Let him remember that, unlike wine, whisky does not

---

8 *This stricture on single malt whiskies appears in a number of books from this period on, but I have been unable to determine where it originated—possibly here. It appears twice in his text and it is surprising to find MacDonald giving space to this suggestion, given his manifest enthusiasm for 'single' whisky.*

'age' in the bottle. Even if he is buying only a few bottles at a time, he can improve a very ordinary spirit almost out of recognition if he buys one of those small sideboard casks of sherry-cask wood which cost forty shillings or thereabouts.[9] He need not grudge the price, for it will put shillings on to the value of every bottle he drinks. As a rule the casks hold three bottles and must be refilled when they are half empty. The selective, ennobling processes of time and absorptive wood can play their part on his sideboard just as well as in the bonded warehouse, and after a few months he will begin to notice an undeniable ripening and mellowing in the whisky. This will tend to grow ever more apparent, for the older whisky will 'doctor' the newer, smoothing out its rough corners and speeding up maturation.

There is, indeed, no reason why a man equipped with even a small cask of this kind should not make some blending experiments on his own account. He may, for example, begin by trying the effects of a mixture of a blend containing a large proportion of Highland malts with one of the common mostly-grain types. Or, growing bolder, he may write to one or two of the better-known Scottish distilleries and ask for a few bottles of a mature single whisky. My

---

[9] *While the cost has increased, the advice remains sound. Small casks can be easily obtained from a number of online suppliers.*

own experience is that he will usually be lucky. With a few bottles of a Speyside, an Islay, and a sound Lowland malt he will be well embarked on his career as a blender, to his own edification and the delight of his friends.

A warning should be uttered here about the age-factor in whisky. The best opinion holds that a whisky which has been more than fifteen years in the cask tends to deteriorate.[10] It becomes 'slimy'—not in every case but frequently enough to make one treat with some degree of critical detachment the blind worship of seniority in whisky.

If the student has more time and means at his disposal than we have hitherto presumed, he may carry out his experiments on a somewhat larger scale and, in the fashion of the Scots gentry of an older day, get him a ten or fourteen gallon cask. This he should fill with a six-to-eight year-old whisky, putting a tap in half way down. When he has drunk down to the tap he will then fill once more with a potable but not too venerable spirit. He will soon find himself the happy owner of some quite unusually good

---

10 *This was for many years the prevailing orthodoxy in the whisky industry. Until relatively recently, twenty-five or at most thirty years was thought the absolute limit for cask maturation. Today, either consumer taste has changed or wood management and the understanding of the underlying chemistry has evolved. Possibly both are true, for it is no longer unusual to see whiskies of fifty or more years of age offered for sale, albeit at stratospheric prices.*

whisky, even if he cannot in a day or two utter the words spoken to Professor Saintsbury by a member of an old Scottish family, 'It should be good. It comes from a hundred-gallon cask which has never been empty for a hundred years'. At the very thought of such riches and antiquity the least impressionable lover feels impelled to doff a respectful if covetous hat. A concise account of the true technique of the care of whisky was given not long ago in the will of a gentleman who left to an 'old and dear friend' a ten-gallon cask of Scotch whisky:

'It is my desire, but without imposing any condition, that he shall from time to time, when the contents of the said cask shall reach the level of the tap now in the middle at the end of the cask, refill the said cask with five gallons of Scotch whisky, so that the said cask shall not from time to time be tilted to draw off the contents.'

I am told that it is now no use saying that there are only two correct ways to drink whisky: neat and in small quantity as a liqueur after a meal (but for that you must have something better than grain spirit), or diluted with plain water. The world asks for whisky—and soda, and the syphon destroys the work of the still. No more effectual way of ruining the flavour of a good whisky could have been imagined than this one of drowning it in a fizzing solution of carbonic acid gas. For people who like the taste of soda-water it is no doubt an excellent combination, though one may ask, why do they add whisky to it? Only the other day I heard someone who ought to have known better say that he found whisky-and-water 'a dead drink' after whisky-and-soda. On the same principle one would be entitled to describe, say, Montrachet as 'a dead drink' after a liaison with cheap champagne.

The Scots recognise one or two additional methods of imbibing whisky and, as few of them destroy the fine flavour of good malt spirit, they may appropriately find mention here. There is, to begin with, toddy, a potent and stimulating compilation too often used as medicine and too rarely as the crown and climax of the festival. Its ingredients are simple and few: whisky (and for this and all other hot whisky drinks the whisky must not be the commonplace spirit of the hoardings, otherwise the fumes may be too unpleasant to be penetrated by the nose), sugar, and boiling water. The last should be poured into a tumbler until it is half full; it should be kept there until the glass is well warmed and then poured out. Loaf-sugar (the quantity depending upon taste and past experience) should now be melted in the tumbler with a wine glassful of boiling water. It should, nay, it must, if sacred traditions are to be observed, be stirred with a silver tea-spoon. Now add a half glassful of whisky; stir. Some more water; another half-glass of the spirit. Stir, and serve hot.[11]

Athole Brose is a beverage of a different character, its constituents being whisky, cold water, and heather honey, though as a matter of fact it is said to have consisted originally of oatmeal and whisky! A formidable association! Put a pound of honey into a bowl and dissolve it in cold water. About a teacupful of water will suffice. Stir—once more with a silver spoon—and when the water and the honey are intimately mixed add one and a half pints of whisky. You must now stir until a froth rises, upon which you bottle and keep tightly corked. On the authority of Meg Dods,[12] the supreme arbiter of the Caledonian kitchen, 'the yolk of an egg is sometimes beat up

---

[11] *Observe the satisfied smile of the gentleman in the Mutter's Bowmore advertisement (page 143). A whisky toddy has clearly worked its wonders. I will eschew discussion of the cocktail recipes, though readers may care to conduct their own experiments.*

[12] *The reference is to the fictional Mrs Margaret Dods, chatelaine of the Cleikum Inn in Sir Walter Scott's novel* St Ronan's Well *(1824 evidently a favourite of the author—see page 111). Inspired by the character's culinary expertise, some two years later Mrs Christian Isobel Johnstone was to produce a cookbook under the title* Meg Dods' Cookery: The Cook and Housewife's Manual by Mistress Margaret Dods. *This proved phenomenally successful, remaining popular for the next hundred years or so!*

Mrs Johnstone was married to Scott's publisher; the opening chapter is reputed to have been written by Scott himself, who cheerfully recognised the production in later editions of his novel; Thomas de Quincey praised her work and Heston Blumenthal has acknowledged the book as the source for his Cucumber Sauce.

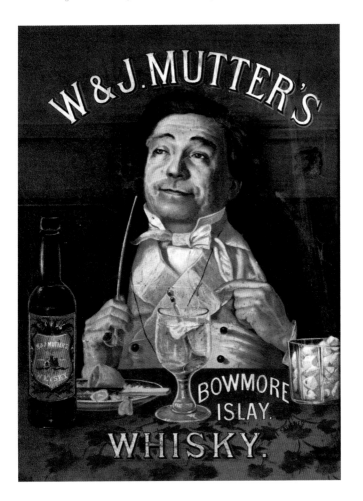

with the brose'. This drink, which has the honour of being mentioned in *The Heart of Midlothian*, is carried into the sergeants' mess of the Argyll and Sutherland Highlanders on Hogmanay, borne by two subalterns and escorted by a piper and all the officers of the regiment. A quaich is filled for every officer and sergeant.

Other ancient whisky recipes are Auld Man's Milk, Highland Cordial, Highland Bitters, Het Pint and Caledonian Liquor. Auld Man's Milk is made by beating the yolks and whites of half a dozen eggs separately. Add to the yolks sugar and a quart of milk (or thin cream), and to this about half a pint of whisky. The whipped whites are then united with this mixture and the whole is gently stirred in a punch-bowl. Flavoured with nutmeg or lemon zest, this makes an admirable morning dram.

Highland Cordial is a somewhat more elaborate decoction whose foundation is a pint of white currants to which are added a bottle of whisky, the thin peel of a lemon, and a teaspoonful of essence of ginger. These are mixed and allowed to stand for forty-eight hours, when the liquid is strained. Then a pound of loaf sugar is added and given a day to dissolve. Now bottle and cork, and, in three months, begin to drink.

Highland Bitters is an extremely ancient beverage

This discerning whisky drinker looks askance at the soda siphon and offers it a cold shoulder—just as MacDonald recommends!

*Perhaps these hardy gentlemen are searching for the perfect blend—
which one presumes would be made by Dewar's*

with a recipe somewhat difficult to assemble on short notice. First an ounce and three-quarters of gentian root and half an ounce of orange peel should be cut into small pieces and bruised in a mortar with an ounce of coriander seed, a quarter-ounce of camomile flower, half an ounce of cloves, and a quarter of an ounce of cinnamon stick. Now put in an earthenware jar and empty two bottles of whisky over it. Keep the jar air-tight for about a fortnight and then strain and bottle.

Het Pint was once—so Scottish legend runs—made with light wine and brandy. But ale and whisky have taken the place of these exotic ingredients in the beverage which was once carried (in a copper kettle) about the streets of Edinburgh and Glasgow on New Year's Morning and was also consumed on the night before a wedding and at a lying-in. A nutmeg is grated into two quarts of mild ale and brought to boiling point. To a little cold ale add some sugar and three well-beaten eggs. This is now slowly mixed with the hot ale, care being taken that the eggs do not curdle. A half pint of whisky being added, the whole is brought to boil again and then briskly poured from one vessel to another until it becomes smooth and bright.

A simpler beverage is Caledonian Liquor, which is made by dropping an ounce of oil of cinnamon on two and a half pounds of bruised loaf sugar; a gallon of whisky—the best you can lay hands on—is added to this and, when the sugar is dissolved, the liquor is filtered and bottled.

For the serious-minded lover of whisky, however, recipes taken from antique household books will have small attraction. All his time will be devoted to that romantic, unending quest of the true participator in the mysteries of aqua vitae—the search for the perfect blend.[13] To some men it has been vouchsafed to put lips to a glass of this legendary liquor—there was one occasion in my own life when I thought that the luck was mine—but never has a man been known to possess a bottle of the peerless distillation at the moment when he is—no, not describing its graces—but faintly adumbrating them by fantastic and far-fetched analogies, apologized for even as they are uttered. Most of us are content to believe that such a whisky exists—must exist—and to go on looking for it. One day before we die some unknown fellow traveller in a railway compartment, some Scots ghillie or Irish rustic, may produce a flask or unlabelled bottle and we shall find ourselves at last in the presence of the god himself, Dionysos Bromios, God of Whisky. And then our sensations, rewarded after years of disappointment and imperfect delight, may share the ecstasy of him who, in C.E. Montague's

---

[13] *A strange contradiction to find the search for the perfect blend concluding a book praising single malt whisky. It exemplifies the pleasure and frustration; the elusive pleasure of reading Aeneas MacDonald—first poet of whisky.*

Another Temple Gone,[14] tasted the whisky of the priestly Tom Farrell:

'Its merely material parts were, it is true, pleasant enough. They seemed while you sipped, to be honey, warm sunshine embedded in amber and topaz, the animating essence of lustrous brown velvet, and some solution of all the mellowest varnish that ever ripened for eye or ear the glow of Dutch landscape or Cremona fiddle. No sooner, however, did this probable sum of all the higher physical embodiments of geniality and ardour enter your frame than a major miracle happened in the domain of the spirit: you suddenly saw that the most freely soaring poetry, all wild graces and quick turns and abrupt calls on your wits, was just the most exact, business-like way of treating the urgent practical concerns of mankind.'

---

14 *Charles Edward Montague (1867–1928) is hardly remembered today and, I daresay, not read at all. But in his day his work was both popular and critically acclaimed, with his interest in moral and philosophical problems compared favourably with that of Joseph Conrad.*

Another Temple Gone *is from* Fiery Particles, *a collection of short stories published in 1923 which remained in print until the early 1950s. Robert Bruce Lockhart mentions it favourably in his* Scotch: The Whisky of Scotland in Fact and Story. *Copies can be found on the web for a few pounds: it is well worth tracking one down to enjoy a superbly written tale by a fine if neglected writer.*

Tom Farrell is an Irish distiller of poteen whose still is detected and destroyed, albeit reluctantly, by Maguire, a sergeant of the Gardai. Later he assists Farrell in escaping justice but not before he has tasted his whiskey. 'Mother of God!' the sergeant exclaimed. 'What sort of hivven's delight is this you've invented for all souls in glory?'—the very lines which preface Whisky.

ACKNOWLEDGEMENTS

A number of people have helped with the Appreciation in this edition of *Whisky*. Fellow whisky writers Dave Broom and Charles MacLean and Dr Nicholas Morgan (Diageo) and Doug Stone assisted in some tricky points of detailed identification, and I am grateful to them for their learned assistance so readily given.

Particular thanks are due to the late Mrs Anne Ettlinger, George Malcolm Thomson's daughter and literary executor, who shared reminiscences of her father, allowed access to his private correspondence, and provided the picture of him.

FURTHER READING

Alistair McCleery
The Porpoise Press 1922–39
Merchiston Publishing, 1988

George McKechnie
George Malcolm Thomson: The Best-Hated Man
Argyll Publishing, 2013

J.A. Nettleton
The Manufacture of Whisky and Plain Spirit
Cornwall & Sons, 1913